The Late Middle English Version of Constantinus Africanus' *Venerabilis Anatomia* in London, Wellcome Library, MS 290 (ff. 1r-41v)

The Late Middle English Version of Constantinus Africanus' *Venerabilis Anatomia* in London, Wellcome Library, MS 290 (ff. 1r-41v)

By

Jesús Romero-Barranco

Cambridge Scholars Publishing

The Late Middle English Version of Constantinus Africanus' *Venerabilis Anatomia* in London, Wellcome Library, MS 290 (ff. 1r-41v)

By Jesús Romero-Barranco

This book first published 2015

Cambridge Scholars Publishing

Lady Stephenson Library, Newcastle upon Tyne, NE6 2PA, UK

British Library Cataloguing in Publication Data
A catalogue record for this book is available from the British Library

ISBN (10): 1-4438-8040-X
ISBN (13): 978-1-4438-8040-4

To Javier Calle-Martín
(Department of English, University of Málaga)

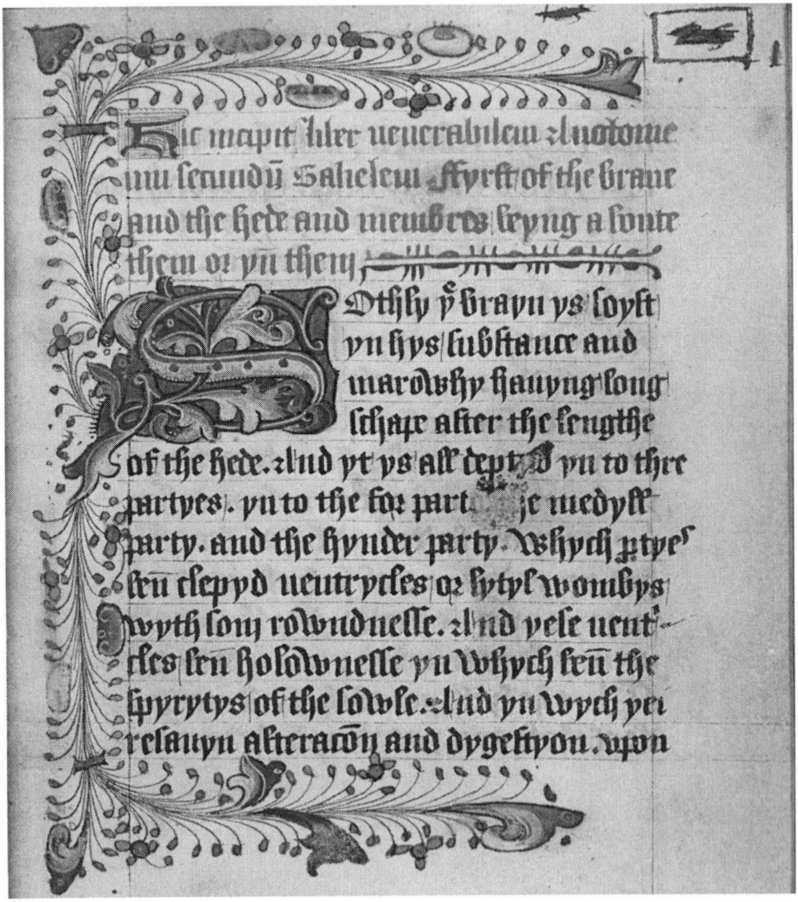

Hic incipit liber venerabilem anotome
um secundū Gahelem. ffyrst of the brane
and the hede and membres beyng a boute
them or yn them

Sothly ye brayn ys soft
yn hys substance and
marowhy hauyng long
schape after the lengthe
of the hede. And yt ys all departed yn to thre
partyes. yn to the for party the medyll
party. and the hynder party. Whych partyes
ben cleped ventrycles or lytyl wombys
wyth som rowndnesse. And yese ventri-
cles ben holownesse yn whych ben the
spyrytys of the sowle. And yn wych yei
resauyn alteracioun and dygestyon. upon

TABLE OF CONTENTS

LIST OF FIGURES

LIST OF TABLES

List of Abbreviations

a.	Adjective
b.	Adverb
bot. marg.	Bottom margin
c.	Conjunction
c.	*Circa*
d.	Determiner
del.	Cancellation
dup.	Repetition
e-MED	Electronic *Middle English Dictionary*
ext. marg.	External margin
f.	Folio
ff.	Folios
LALME	*Linguistic Atlas of Late Middle English*
ME	Middle English
MED	*Middle English Dictionary*
n.	Noun
ng.	Deverbal noun
OED	*Oxford English Dictionary*
p.	Preposition
pl.	Plural
r.	Recto
sg.	Singular
v.	Verso
v.	Verb
va.	Anomalous verb
W290	London, Wellcome Library, MS Wellcome 290 (ff. 1r-41v)

ACKNOWLEDGEMENTS

The present research has been funded by the Spanish Ministry of Science and Innovation (grant number FFI2011-26492) and by the Autonomous Government of Andalusia (grant number P11-HUM7597). These grants are hereby gratefully acknowledged.

Dr. Javier Calle-Martín deserves sincerest gratitude for having introduced me to such a fascinating field of study and for his guidance and good advice over the past years.

I am particularly indebted to Amanda Engineer and Natalie Walters (Archivists, Department of Archives and Manuscripts, The Wellcome Library), Natalie Costaras and Crestina Forcina (Picture Researchers, Historical Collection, The Wellcome Library) and Dr. Christopher Hilton (Senior Archivist, The Wellcome Library) and to the Wellcome Trust for granting us permission to use this unique collection of manuscripts.

Special thanks are also due to Dr. Antonio Miranda-García and Joaquín Garrido-Garrido (computer technician), who designed the *ad hoc* tool for the automatic retrieval and handling of linguistic data.

FOREWORD

I was introduced to Mr. Romero-Barranco in Córdoba in December 2013 at the XXV International Conference of the Spanish Society for Medieval English Language and Literature *(Selim)* by Dr. Javier Calle-Martín and there attended their paper "On the Use and Distribution of Object Clause Links in Early English Medical Writing." As a result of such an interesting episode I started following Mr. Romero-Barranco's academic tracks, which include my reading of some of his publications, such as 2014 "The Split Infinitive in the Asian Varieties of English" (*Nordic Journal of English Studies* 13.1: 129-146); and 2014 "On the Use of that/zero as Object Clause Links in Early English Medical Writing" (*Studia Neophilologica* 86.1: 1-16). More recently the interested reader can easily access 2015's "On the Use of *make to* vs. *make ø* in Early English Medical Writing" *(Atlantis: Journal of the Spanish Association of Anglo-American Studies* 37.1: 157-177). Despite the fact that attending academic conferences tends to be entertaining, it was also fruitful on such an occasion.

If we also take into account that the author of this book finished his master's degree in 2013, and that two interesting publications dealing with texts preserved in MSS Wellcome 411 and 290 *(Analecta Malacitana* numbers 35 and 36, both from 2013) came from his scholarly efforts, one is rather tempted to have great expectations about the future of philological and medieval studies in Málaga. These may even pervade to other less fortunate locations in Spain and maybe abroad and overseas.

So when I received a call to duty from Málaga in order to provide some introductory words for the publication of *The Late Middle English Version of Constantinus Africanus'* Venerabilis Anatomia *in London, Wellcome Library, MS 290 (ff. 1r-41v)* I started pondering on how could I make use of this text's very interesting lexicon and rake "the brayn of Nucha" (line 830, folio 27 recto) without becoming too involved with Arabic *nuha (nuẖā' or nuẖāʕ),* and such like (because this text shows an unusually strong Arabic influence in its vocabulary),[1] how to sound convincing enough without having to resort to the efforts that may trigger "the goyng oute of the ȝerbus . of the gutt . to the ballok codde" (ll. 897-898, folio 29 verso) something not unusual in those slightly elderly academics that lead sedate parochial lives in the vicinity of the countryside and its not unseemingly beastly joys, and how to say something approximately as convenient to the case now commented on as "þ*er* the cotys of the ballok and the skyn growen or bene brede . yn whych the stonys be contenede" (927-929, folio 30 verso) which might not at the same time provoke uncomfortable surfeits of verbosity and eternal damnation for having promoted unnecessary profanity in the academic discourse. Let me explain just a little further why I have chosen those three examples from the *Anatomy.*

Late Middle English texts seem to be closer to the modern world than earlier medieval works because their spelling – despite the mannerisms and oddities- is closer to the 15th and 16th more extended standardised forms established by the Chancery and the Printers than the previous Early Middle English or Old English texts which have come down to us. Their phonology, morphology and partially their syntax and lexicon make them more remote. The three short sentential examples I chose may be proof of that. However they are also proof that the readability and interest of this medical treatise is particularly good in idiomatic terms: most present day speakers of English would most likely word these expressions similarly in the familiar (or everyday) register –quite a world apart from the physicians' jargon. This is fairly unusual in scientific prose from the Middle Ages which tends to make contents even more remote from the modern reader by sticking closely to Medieval Latin syntactic and lexical patterns even when the works are not mere translations from foreign texts.

Constantinus Africanus' *Venerabilis Anatomia* runs for 1290 lines, about ten thousand words, in manuscript Wellcome 290. It constitutes a sufficient corpus for the sort of studies that this edition provides and it is also a remarkable example to illustrate that the vernacular simplicity of the rhetorical construct of this work is not at all at odds with the precise medical terminology and technical knowledge displayed by its author(s). I agree that reading fifteenth-century English medical treatises is not everybody's cup of tea, but I must also emphasize that going through this text in some detail and reading the edition and studies that precede it has stirred a new interest in my approaches to medieval philology. I hope that such an interest may also provide other academics with an example of what can and needs to be done with Late Middle English scientific prose for the advancement of the disciplines involved in materializing how we still need to learn from the past to have a brighter future. Disciplines that blend together the CA (computer age) tools, techniques and methods with the BC (before the computer) paradigms in the way in which the Málaga research groups tend to do and Mr. Romero-Barranco has wisely demonstrated that can be done.

S. G. Fernández-Corugedo
University of Oviedo

Notes

[1] The influence of Arabic words in medieval English (despite the moderate number that OED may list as neat borrowings) is salient in the fields of mathematics, astronomy, philosophy, literary theory, astrology, gardening and agriculture, chemistry, war, games... See for instance: Charles Burnett 1997: *The Introduction of Arabic Learning into England.* The Panizzi Lectures. London: British Library and Jessica Wilson 2001: Arabic in Middle English. http://homes.chass.utoronto.ca/~cpercy/courses/6361Wilson.htm. More specifically: Mary S. Serjeantson 1961 (1935): *A History of Foreign Words in English.* London: Routledge and Kegan Paul. [Chapter ten is devoted to Arabic loanwords (and those from other Asian languages).]; and Andrew Breeze 1991: Old English Ealfara, 'Pack-Horse': A Spanish-Arabic Loanword. *Notes and Queries* March (1991): 15-17. Mentioning Chaucer's *Treatise on the astrolabe* may also be to the point here, in the same way that Constantinus Afer is mentioned in the *General Prologue* (lines 430-434) as "Constantyn the cursed monk". The field of medicine, which has received some attention (see M. A. R .Al-Fallouji 2009: Arabic Influence on English Language in Medicine and Routes for the Linguistic Transfer. Find it at http://www.ihams.org/index.php; and Erin Ellerbeck 2005: Middle English Medical Terminology: The Body. http://homes.chass.utoronto.ca/~cpercy/courses/6361ellerbeck.htm) is still in need of a more in-depth study (maybe to appear as a result of Juhani Norri's project entitled *Dictionary of Medical Vocabulary in English, 1375–1550* (http://www.uta.fi/ltl/plural/common/projects/dictionary_medical_vocabulary.html), as this text of *Venerabilis Anatomia* clearly demonstrates.

CHAPTER ONE

THE MANUSCRIPT

The present chapter focuses on the manuscript witness, paying attention to the source and contents of the volume, together with a codicological and palaeographic analysis. In addition, a full folio of Wellcome 290 (ff. 1r-41v) is reproduced so as to provide the reader with an original sample of the manuscript under study.

1.1. Source and Contents

The heterogeneous nature of Middle English was represented by the so-called Middle English dialect[1], which co-occurred in England, particularly in the period 1100-1500. Extant texts from this period have come to us through handwritten documents, which cover a wide range of categories, such as scientific writing, literature or legal proceedings, among others. From all these categories, scientific writing could be highlighted, as it offers a more natural representation of the language, far from the usually ornamented literary or legal language. Apart from the vernacular texts which may have served the ordinary layman, there were others that were made for the use of professional physicians, among which versions of Hippocrates or Galen can be found (Talbot 1967: 191).

The manuscript under scrutiny in the present edition is housed in London, Wellcome Library. It is referenced as MS Wellcome 290, and entitled *Pseudo-Galen, Claudius, 131 – 201*, comprising 56 folios of which the last three are blank (Moorat 1962: 186). The present edition, however, focuses on the first part of the volume (ff. 1r-41v)[2], housing an anonymous Middle English translation of Constantinus Africanus' *Venerabilis Anatomia* (henceforth W290).

As for the historical background of Constantinus Africanus, he is supposed to have been born in the 11th century in modern Tunis or Sicily, "hence his agnomen or title of place, Africanus" (de Weever 1988: 95). He was probably educated in the famous academy patronized by the Fatimid rulers in Cairo and he seems to have been driven out by the invading nomads when he returned to his native country (Newton 1994: 19). He was certainly an important character when he first arrived in Salerno, where he was considered a respected figure and with “some considerable worldly backing” (Newton 1994: 20). As de Weever puts it,

> Constantinus’ translation of Arabic medical texts into Latin gave the West a number of important works. These formed the foundations of modern science and biology. He was a much cited authority from the twelfth until the sixteenth century, and his translations were widely circulated (1988: 95)

An illustration of the importance held by Constantinus Africanus in the field of mediaeval medicine, where he was considered an authority, is a quotation from Chaucer´s *General Prologue* of *The Canterbury Tales*,

> With us ther was a doctour of phisik; In al this world ne was the noon hym lik, To speke of phisik and of surgerye For he was grounded in astronomye. […] Wel knew he the olde Esculapius, And Deyscorides, and eek Rufus, Olde Ypocras, Haly, and Galyen, Serapion, Razis, and Avycen, Averrois, Damascien, and Constantyn, Bernard, and Gatesden, and Gilbertyn (Chaucer, *General Prologue* of *The Canterbury Tales*, ll. 429-434)

In this fragment, Chaucer is introducing one of the pilgrims who was part of the company, the ‘doctour of phisik’, and, in order to give an account of his mastery on the field of Medicine, Chaucer assures that he is well acquainted with the works of Constantinus Africanus, among others. This quotation by such an influential author as Chaucer perfectly portrays the significance of Constantinus Africanus in the period, as Chaucer places him next to other prestigious scholars, namely Galen or Avicenna.

The proliferation of Middle English translations of this scientific material in Latin shows the interest of mediaeval England in this kind of literature and, besides, the emergence of the English language as the language of science, far from the colloquial registers to which it had been restricted after the Norman Conquest (Blake 1992: 5).[3] According to Voigts and Kurtz (2000), there is another copy of MS Wellcome 290 in the Wellcome Library, referenced MS Wellcome 397 (ff.1r-14v), which is incomplete.[4]This copy of MS Wellcome 290 (ff. 1r-41v) is particularly important as it contains a prologue attributing Lanfranc of Milan as the main authority of the text, a fact that suggests that the treatise was originally composed in Latin by Constantinus Africanus and, some two centuries later, Lanfranc of Milan was somewhat involved in that particular witness. He was a native of Milan who was a student of the great Italian surgeon William of Saliceto and he was responsible for introducing the knowledge of Islamic and Italian writers to northern Europe (Keiser 1998: 3645). He eventually became a popular professor of surgery at the Collège de St. Côme[5] in France, where "the dean of the faculty requested that he wrote about what he knew, and Lanfranc put on paper his knowledge of anatomy, embryology, ulcers, fistulae, fractures, and dislocated joints, as well as some nonsurgical subjects such as herbal medicines: the result became a major work on medicine" (Kelly 2009: 67). This work was carried out in two stages; first, after he moved from Milan to Lyon, the physcian wrote the *Chirurgia Parva* ('Short Surgery'), a surgical summary for the education of his sons; some years later, he moved to Paris, where he joined the surgeons' guild and composed *Chirurgia Magna* ('Complete, or Great, Surgery'), which is composed of five treatises[6] (Prioreschi 1996: 458):

1) Definition of Surgery, qualities of a surgeon, anatomy, wounds and ulcers.
2) Wounds on particular parts of the body with their anatomy.
3) Surgical diseases *a capite ad calcem*[7].
4) Fractures and dislocations (*algebra*).
5) Drugs.

1.2. Codicology

This section analyses the volume from a codicological point of view. Consequently, features such as material, dimension, ink, decoration, quiring, ruling and foliation will be faithfully described. These descriptions are the result of the examination of the digitized images provided by the Wellcome Library, together with the accurate information provided by Moorat's *Catalogue of Western Manuscripts on Medicine and Science in the Wellcome Historical Medical Library* (1962).

1.2.1. Material and Dimension

The volume is written in vellum[8] and bounded with 19th century vellum binding (Moorat 1962: 186). It can be safely said that it is in overall good condition, although some damage can be observed on the margins, which are slightly cropped. The dimensions of the volume are 18 x 13 cm (Moorat 1962: 186).

1.2.2. Decoration and ink

The decoration is an essential part of the manuscript book and has historically attracted the attention of scholars. In fact, many manuscripts include rich illuminated borders and miniatures, therefore being unusual for a completed medieval book to comprise nothing but plain regular script (De Hamel 1992: 45). In the Middle Ages, during the late Middle English in particular, "the non-illustrative decorative elements, which have been studied so intensely for earlier periods, have generally been neglected in art historical research" (Derolez 2003: 39-40). The reason for this lies in the fact that innovation took place in the pictures that accompanied the text, while initials and other elements of decoration were more and more standarized due to production needs.

W290 is written in two main colours, brown and red, the former used for the body of the manuscript while the latter is preferred for the headings and the endings of the chapters (usually rendered in Latin). The brownish colour used in the body slightly fades[9] after f. 25, and it is kept in this shade until the end of the volume.

The scribe also represents large illuminated initials in gold and colours with feather ornaments at the beginning of each chapter (ff. 1r, 29r, 37r and 41v). Furthermore, some initials are illuminated in gold and blue in order to highlight important sections within a chapter. Paragraph marks are represented in an unusual form, in gold and blue on a red ground (Moorat 1962: 186).

1.2.3. Quiring

The quiring of this volume employs an octavo opening (Moorat 1962: 186). This was a regular practice during the Middle Ages, although "the relative thinness of the material often induced producers of books to use quires of more than four *bifolia*, indeed of six or up to twelve and even more *bifolia*" (Derolez 2003: 32). W290 presents a regular quire, which was not very usual in the late Middle Ages, as books often included single folios in between the pages of the quires. Seven quires were needed to create this volume, which account up to 56 folios.

1.2.4. Ruling

Ruling was used to keep the lines of the text straight. According to Petti, "before the writing commenced [...] a frame was provided for the writing area of each page and the lines ruled" (1977: 6)[10]. In the late 15th century ruling became less fashionable and only the frame remained, an element which would be omitted from the 16th century on (Petti 1977: 6). Constantinus Africanus' *Venerabilis Anatomia* is written in one single column, presenting both the frame (prolonged to the outer edges) and the lines ruled, suggesting that this witness was plausibly written before the middle of the 15th century (see Plate I). In fact, Derolez states that "two-column rulings were preferred throughout late Middle English and early Modern English, but in the fifteenth century a renewed preference among some for layouts with long lines is detectable, probably under Italian Humanistic influence" (2003: 37).

1.2.5. Foliation

Foliation was not a usual feature until the end of the Middle Ages, although a sort of numbering or sequencing may be found. Thus, *quire-marks*, consisting of a Roman numeral, were written in the lower margin of the first or last page of a quire. In addition, *signatures* could be found, indicating not only the order of the quires, but also of the *bifolia* of each quire (Derolez 2003: 35). W290, however, does not follow any of these two practices, as folios are systematically numbered with arabic numerals at the top right corner of each folio recto, sometimes blurred by the passing of time, as shown below.

Fig. 1.1. Foliation in W290 (f. 34r)

1.3. Palaeography

Palaeography could be defined as the study of the different scripts with the objective of providing an estimate date of composition. Denholm-Young states that "the business of a palaeographer is not only to read, classify, date, and determine the provenance of a manuscript, but to recognize textual errors that spring from the scribe´s misreading of what he is copying" (1964: 1). The aim of the present section is then to describe the most relevant letterforms of W290 in order to propose an approximate date of composition for the manuscript. Features such as numerals, marginalia, catchwords, manuscript corrections and punctuation are also analysed.

1.3.1. Script

W290 is written using a fairly legible hybrid script, consisting of a combination of bastard Anglicana with some characteristic features of the Secretary script.

According to Petti, the Anglicana script "first appeared in England in the 12th century" and it "predominated until close on the middle of the 15th century" (1977: 14). In these three centuries the style went through some alterations, although some characteristics predominated over time, such as the double-lobed <a>, the <d> with a looped stem, a two compartment figure 8 form of the letter <g>, the long forked <r> and the <w> made up of two looped l´s and a *3* (Petti 1977: 14).

It was in the middle of the 15th century that another script appeared, the Secretary script, which grew in popularity becoming widespread towards the end of that same century. Inevitably, the coexistence of these two scrips made them borrow "from one another both in features of general style and in use of graphs" (Petti 1977: 15). Among the distinctive letters of the Secretary script are the single-lobed <a> with a pointed head; the single compartment <g> with a pointed head and a small tail; the short, right-shouldered and *v* form of <r>; the final <s> looking like a small *B* or *c* and *3* run together; and the <w> usually resembling double *v* (Petti 1977: 14).

Bastard Anglicana is the result of "the borrowings among the two different scripts. Its letterforms are defined as larger than those of Anglicana Formata, better spaced and with greater emphasis placed upon its calligraphic execution, as a result of the merging of two previous variants of the Gothic script, anglicana and textualis" (Parkes 1969: xviii; Brown 1990: 81; Derolez 2003: 140). Characteristic features of this blended script are the angularity and squarish proportions of the letterforms, together with a productive ease of writing.

The script in W290 combines a bastard Anglicana hand with some distinctive shapes from the Secretary script. Actually, Roberts (2005: 164) and Brown (1990: 100) term this kind of script as *cursive anglicana formata hybrida* and *anglicana formata hybrida*, respectively. Similar instances of this hand can be found in Parkes (1969: pls. 7-8), Brown (1990: pl. 37) or Derolez (2003: pls. 84-86).

Fig. 1.2. Inventory of letterforms in W290

The clear evidence of this coexistence is the fact that W290 presents different renderings for the same letter in specific cases, where the use of one graph or the other has to do with its position within the word. Figure 1.8 above reproduces the inventory of minuscules used by the scribe where letters have been accordingly numbered for reference purposes.

Among the letters following the Anglicana style, the following stand out: the two-lobed <a> (2); the <h> with a long supralinear stroke curving right and forming a lobe, with a second infralinear leg which curves slightly to the left (9); the letter <k> with a long supralinear stroke slightly curving right, from which two small strokes stem out upward and downward, the upper one forming a lobe occasionally (12); the long <s> (21) and the sigma-like <s> (22); and the <x> with two crossed legs with a small horizontal stroke (27).

The Secretary script, on the other hand, is featured by: the single-lobed <a> (1); the letter <b> with two rounded lobes, the lower of which is slightly bigger than the upper one (3); the <d> with a lobe and a curve ascender (5); the <g> consisting of a single compartment with a rounded head and a small tail (8); a vertical stroke with a small rightward lobe at the top for <l> (13); an infralinear stroke with a small leftward lobe at the top for <q> (18); the right-shouldered <r> (19) and the two-shaped <r> (20); and the canonical Secretary form for <w> (26).

All in all, the palaeographic analysis suggests that the manuscript was composed towards the middle of the 15th century.

1.3.2. Numerals

According to Hector, roman numerals were dominant in England until the 16th century, when they were replaced by arabic numerals. In this vein,

Hector states that “they are not really common in English documents before about 1530” (1958: 43; also Petti 1977: 28). Roman cardinal numbers are observed in the manuscrpt under study, where the last number in a group is found to be represented using the *i-longa* <j>, as in <ij>, <iij> or <iiij>, representing two, three and four, respectively. Furthermore, whenever the Roman numeral 5 is rendered (fig. 1.11), it is always represented using <v> instead of <u>.

Fig. 1.3. Two

Fig. 1.4. Three

Fig. 1.5. Seven

Ordinal numbers, in turn, are also reproduced in Roman figures, but -th and –de are added in superscript, the raised letter being there “from habit rather than for indicating omission” (Petti 1977: 24).

Fig. 1.6. Fourth

Fig. 1.7. Sixth

Roman numerals in W290 also point to the fact that the treatise was translated towards the mid-fifteenth century, thus corroborating the conclusions obtained in the previous section.

1.3.3. Marginalia

Apart from the numbering of the folios, the external margins serve as a kind of index with which the reader could identify the key topics in the different parts of the treatise (ff. 2v, 3r, 4r, 9v, 33v and 35r), consisting mostly of anatomical terms (Moorat 1962: 186).

From a palaeographical point of view, these annotations seem to have been written by a 16th century hand (Moorat 1962: 186), following the conventions of the Secretary Elizabethan script. According to Petti, this script was “a compact hand, written with a very fine nib held at a slightly oblique angle. It was almost entirely cursive except in its set form, and could obviously be written at great speed" (1977: 17). Furthermore, scribes felt free to introduce individual variations into their writings, to the point that some of the letters were almost unrecognizable (Tannenbaum 1930: 22).

1.3.4. Catchwords

Catchwords were a tool used by scribes in order not to make mistakes in the sequencing of the quires of the manuscript, thus, "the first word of a quire was written at the bottom of the last leaf of the preceding quire, to ensure that the pages of the quire were in the right order" (Petti 1977: 6). In W290, catchwords can be found in ff. 8v, 32v and 40v. In addition, the scribe occasionally writes the number of the chapter starting in the next folio, as in ff. 16v and 24v.

1.3.5. Abbreviations

As a fifteenth-century witness of scientific *Fachprosa*, W290 features a great number of abbreviations. The main function of these abbreviations, as Petti points out, "is to save time and space [...] and [...] make the maximum use of the relatively expensive writing surface" (1977: 22). From a chronological point of view, English documents of the 12th century display the [abbreviation] system in the most elaborate form, while in the latter Middle Ages some of them were gradually discarded (Hector 1958: 29). In addition, Derolez states that genre is found to play an important role in the use of abbreviations, as

> scholastic manuscripts and those of the thirteenth and fourteenth centuries in general contain without doubt the largest number of abbreviations (mostly specific to the subject concerned: theology, philosophy, law, natural science, medicine...), but the degree of abbreviation is far less in liturgical and literary manuscripts (2003: 187).

In spite of being a common mediaeval practice, it is often difficult for the contemporary linguist or palaeographer to compile a complete list of the abbreviations in use during the Middle Ages, as "not all the penmen used the same abbreviations and those abbreviations may stand for different words or letters" (Tannenbaum 1930: 119). As for the typology of abbreviations, different kinds or techniques could be found in mediaeval manuscripts, where Tannenbaum distinguishes contraction, elision, absorption, curtailment, brevigraphs and superior letters, apart from likely combinations of these (1930: 119).

In the present edition, abbreviations have been expanded in italics on the basis of the most frequent spelling of the word appearing in the text. However, when the full form is not recorded in the text, the form appearing in the *MED* is represented. In W290, the scribe makes use of three kinds of abbreviation: *contraction*, *brevigraphs* and *superior letters*.

1.3.5.1. Contractions

Contraction is the commonest method of abbreviation, and consisted in the omission of one or more letters from the middle of a word (Tannenbaum 1930:119; Petti 1977: 22). In W290 the scribe makes use of the tilde with this purpose.

Contractions are mainly used as a substitute for the consonants <m> and <n>, and for the vowels <e> and <i>. The choosing of one letter or the other will depend on the context in which the word appears.

Fig. 1.8. 'dysposyc*i*on' (f. 8v)

Fig. 1.9 'secu*n*de' (f. 6r)

Fig. 1.10. 'co*m*e' (f. 2r)

Furthermore, this kind of abbreviation can be used to indicate the doubling of a letter. In this case, the abbreviated letter has been expanded in italics.

Fig. 1.11. 're*n*nyng' (f. 9v)

Fig. 1.12. 'co*m*mone' (f. 3v)

1.3.5.2. Brevigraphs

Brevigraphs generally represent at least two letters or one syllable, and might resemble one of the omitted letters or be apparently arbitrary in shape (Petti 1977: 23). According to Tannenbaum, "scriveners had developed the habit of contracting certain frequently recurring syllables, both at the beginning and at the end of words" (1930: 125). The following brevigraphs have been observed in W290:

a) The group <per> is abbreviated with one dot on each side of the stem of the letter <p>.

Fig. 1.13. ‘p*er*tyes’ (f. 9r)

b) The groups <pro> and <pre> are abbreviated with letter <p> with a loop:

Fig. 1.14. ‘p*ro*p*re*lyche’ (f. 31v)

Fig. 1.15. ‘p*ro*fitable’ (f. 17r)

c) The group <ra> is represented with a supralinear tilde:

Fig. 1.16. ‘b*ra*nchyng’ (f. 28r)

d) The groups <er> and <re> are abbreviated placing a curved stroke over the previous letter.

Fig. 1.17. ‘man*er*’ (f. 20v)

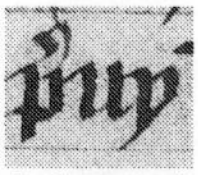

Fig. 1.18. ‘p*re*uy’ (f. 14r)

e) The group <us> was abbreviated with a symbol resembling a <9>[11] with an infralinear stroke curved leftward. This form of abbreviation is only found in the Latin words, as “the occurrence of this final or medial *us*-brevigraph in the vernacular is not very frequent” (Tannenbaum 1930: 127).

Fig. 1.19. ‘Albugeni*us*’ (f. 8v)

f) A version of the Tironian nota, represents the copulative conjunction *and* (Petti 1977: 23), which is rendered as a sort of small *t* on the line with a short horizontal bisector (Tannenbaum 1930: 132).

Fig. 1.20. '*and*' (f. 2r)

1.3.5.3 Superior letters

Superior letters are also a form of abbreviation, although they sometimes seem to be a habit of the scribe. These letters appear both in medial and in final position. When a supralinear letter does not involve an abbreviation, it is simply lowered to the line in the edition. The following superior letters functioning as abbreviations are found in W290:

Fig. 1.21. 'þ*a*t' (f. 11v)

Fig. 1.22. 'w*ith*' (f. 5r)

Among the scribal habits that involve the using of superior letters, the following have been witnessed:

Fig. 1.23. 'þe' (f. 36v)

Fig. 1.24. 'iiijth' (f. 10r)

1.3.6. Manuscript corrections

Professional scribes used to revise their own compositions so that they would detect and correct any scribal error. These corrections could be carried out using different techniques: *erasure*, *cancellation*, *expunction*, *underscoring*, *obliteration*, *vacation*, *dissolution*, *alteration* and *insertion* (Petti 1977: 29). W290 shows evidence of *cancellation,* which consists in striking with ink with one or more straight lines, as shown in the following instances:

Fig. 1.25. ~~elepyd~~ (f.5r)

Fig. 1.26. ~~bone~~ (f. 17r)

Fig. 1.27. ~~lyuer~~ (23v)

Fig. 1.28. ~~goyng~~ (28v)

Fig. 1.29. ~~And that the rubbyng~~ (f. 30r)

Fig. 1.30. ~~ended~~ (35r)

As can be observed above, *cancellation* could delete not only a single word or letter, but also a whole clause or sentence. The scarce number of emendations in W290 constitutes a reason to think that this treatise was presumably a valuable copy, a fact which is further corroborated by framed ruling or conventional margins.

1.3.7. Punctuation

Punctuation in Middle English manuscripts was far less consistent than it is nowadays, and perhaps its apparent arbitrariness could be signalled as the reason why punctuation has been traditionally disregarded in the literature (Jekinson 1926: 154; Denholm-Young 1964: 77; Arakelian 1975: 614-615; Calle-Martín and Miranda-García 2012: 32). Concerning the typology of punctuation symbols in the Middle Ages, Parkes distinguishes the following punctuation marks (1992: 42-45): *punctus*, *punctus elevatus*, *punctus interrogatiuus*, *litteræ notabiliores*, *virgule*, *paragraph mark* and *positura*. In addition to this repertory, it should be taken into account that "notwithstanding the currency of these punctuation symbols in the period, every scribe is ultimately free to imprint his own repertory of symbols" (Calle-Martín and Miranda-García 2012: 32). W290, however, just features two punctuation marks: the *punctus* and the *paragraph mark*.

The period is the most frequent punctuation mark in the witness, amounting up to 914 occurrences. It is employed as a general pause mark to denote sentential and clausal relations (Calle-Martín and Miranda-García 2005: 37-42). Among those usages, the punctus may be used to signal the end of a meaningful statement and the beginning of a new one:

> Fyrst of the brane and the hede and membres beyng aboute them or yn*e* them . Sothly þe brayn ys soyft yn hys substance and marowhy hauyng long schape after the lengthe of the hede . (f. 1r).

To introduce coordinate clauses: the period is employed to link coordinate sentences introduced by the coordinators *and*, *but* and *nor*:

> Toward the share ys set þe bledd*er* . whych ys þe vassyl of the vryn . and yt ys synewy . and the neke of yt ys fleshy os oft nost . and she hath yn hyr two cots the whych ben*e* ij . skynnes . and þ*er* be yn her many smale veynys and art*er*ies . (f. 35r).

> And the lung ys sett yn the same holounesse . but aft*er* hys beyng he loweth to the ryght syde . and yn the myddys of hym he boueth much to the left syde . (f. 26v).

> and on the vpper ende he hath cou*er*yng clepyd Epigloton þat þe tyme of etyng the mete entre not in hem . ne nothyng hurtyng but only eyre . or onythy*ng* lyke to eyre . (f. 15r).

To make suggestions or recommendations to the reader:

> yf we make departyng bytwen*e* þe rall and the adiutorye . wet thow wele þat all þe vaynes þat cu*m*myn from þe lyu*er* abouyn*e* ben*e* dep*er*tyd fro the lesse branche of the gret vayn þat co*m*myth þat cu*m*myth fro the lyuer . and þat vayne ys deuydyd into iij branches . (f. 24r).

To introduce subordinate sentences, the period employed irrespective of the kind of subordination:

> Under the hede yn that place yn þe hynd*er* party ys . Nucha . vnder whych ben*e* ordeyned vij spondels . whych ben*e* vij Ioyntys of the neke . (f. 12v).

> and yt ys departyng of the noryshyng membrys fro þe spyritual . and yt ys clepyd diafragma . þ*a*t ys i*n* englych the myddref . and he holdyth hys place lyke the schape of a bely yn that worchyng . (f. 33r).

> þe cordys mouyng þe arme be de-partyd aft*er* diuers partys . And yn the arme ys anoþ*er* open beyng wythoute þe arme . *and* be spredyth þe arme aboute . of whom ben*e* dep*er*ted diuersse stryng mouyng þe fyngurs (f. 21r).

To introduce a non-finite form of the verb, the period not only marking the beginning of a new clause but also "connecting the *-ing* form with its corresponding subject, which is likely to appear somewhere before" (Calle-Martín and Miranda-García 2008: 365):

> Sothly the art*er*yes þat co*m*men to the forsayd skyn and to the brest . co*m*men by b*ra*nchyng of a branche . dep*er*tyng from a grete arterye . the whych growyth in the lyfte ere . of the herte . (f. 28r).

To introduce a sequential marker:

> by the kuttyng or p*er*syng yn suche placis co*m*meth noysaunce or hurtyng to the dura mater of the brayne . Therfore be þer nor worchyng wyth yren*e* in such places yf yt be possible to be eschuyd . (f. 4r).

> hauyng her begynny*ng* of Nucha . or of the spondels of the brest . wherfore yt schewyth opynly how and yn what maner þat þe apostumes of þese places schulde be kut . (f. 28v).

Apart from all these sentential functions of the period, it may be also used with clausal functions, where the period is employed to indicate the presence of an anatomical term:

> and ys fastned wyth the throte w*ith* the pype of longes clepyd . Trachea arteria . whych ys made of gristles of the vjth payer of synews of the brayn . and yt ys greyny . and hath lytyl hoopes yn the vtter party of the fastnyng wyth the meryngg of the pype of the stomake . (f. 15r).

A numeral:

> Ther ben þen xij . rybbys whych ben fastened w*ith* xij spondelys . and þese spondelys ben clepyd p*ro*p*er*ly the spondys of the brest . the remnaunt of the spondels that ben . v . ben clepyd the spondels of the reynes . whych . xij . rybbys ben bowed yn the maner of a halfe compas . of the whych . vij . of the hynd*er* endys ben fastned wyth the spondelys of the bake . (f. 25v).

To add extra information within the clause:

> the skynne . koue*r*yng the skul . and thys skynne ys made harde . and ys hauyng hymselfe thynne . lytyl flech . not wele sygthty . (f. 6v).

In addition to the sentential and the clausal functions of the period, it is also sporadically used at phrase level, to express complements of nouns:

> And forthi þe wondys of þis place be ful dredeful . and vncurable for þe most p*er*tye . for the ende . of þe vttermore partyes of þe veynes . (f. 40v).

To separate two elements within the same phrase:

> vnder the Iyen in the forparty and vnd*er* the nose ys the chanel . whych ys made of . xiiij . bones . of whych the Ionyng and makyng ys hydde . and not wele knowen*e* . (f. 10r).

The paragraph mark, represented as ¶, appears 79 times in the text. Its use is many a time restricted to paragraph and chapter headings (Petti 1977: 27), marking a key section or topic within the treatise:

> the branche þat ys the iijde . of the vnknown veynys yn the ryght ere of the forsayd hert . ¶ Sothly the art*er*yes þat co*m*men to the forsayd skyn and to the brest . co*m*men by b*ra*nchyng of a branche . dep*er*tyng from a grete arterye . (f. 28r).

Interestingly enough, there are some instances of the paragraph mark in which it acquires some of the functions of the period. Thus, it is occasionally used to introduce coordinate sentences:

> byndyng of the whych growyth oute a synewe bodylyk to the byndyng of a rope . Vnder whych ben*e* many arteryes and venys after thys gretnes askyth . ¶ And thys membre synew or arteryousse so Imayd . ys clepyd a mannys
> yarde . In the hede of whych ys fleshly fulfylly*ng* . (f. 30r).

> Ne thys vpper or herder pannycle ys not conteyned wyth the bone of the brayn panne . lest yt take co*m*mone hurtyng for þe herdnesse of hyt . ¶ But wyth som ioynters medelyng of the skul . by whych the clothy bynedyng . (f. 3v).

To introduce subordinate sentences:

> And benethe þese bones anenste the vpper chawel ys a grete hole . whych ys clepyd peterosum . and yt ys to sey ful herde and stony . whych ys the substance of the bone . mendosa . ¶ By whych passyth the synewe of heryng . here ben therfore nowmbred . vj bonys comp*re*hendyng the brayne . (f. 5v).

To introduce a sequential marker:

> whan þese synews entren the makyng of the muscles mouy*ng* þese p*er*tyes . ¶ Therfore yt accordyth . þat the kuttyng þat be don*e* or owen to be done yn the place of the vpper chanel . *and* of the nose . (f. 10v).

> to the vessyl of yt by swellyng of the ȝarde . castyng or mouyng of the skyn vpon the hede of the ȝerde . he castyth out bett*er* the sperm . ¶ Ferthermore from the Cyphac gone down ij pypys . (f. 30r).

To make suggestions or recommendations to the reader:

> Therfore be þer nor worchyng wyth yren*e* in such places yf yt be possible to be eschuyd . ¶ Ferthermore wete thou wele that the brayn yn the formar p*er*tye vnder þe bone of the forhede hath ij . addicyons lyke hedys of bygg yn the whych þe worchyng and þe vertue of smellyng ys fulfyllyd . (f. 4r).

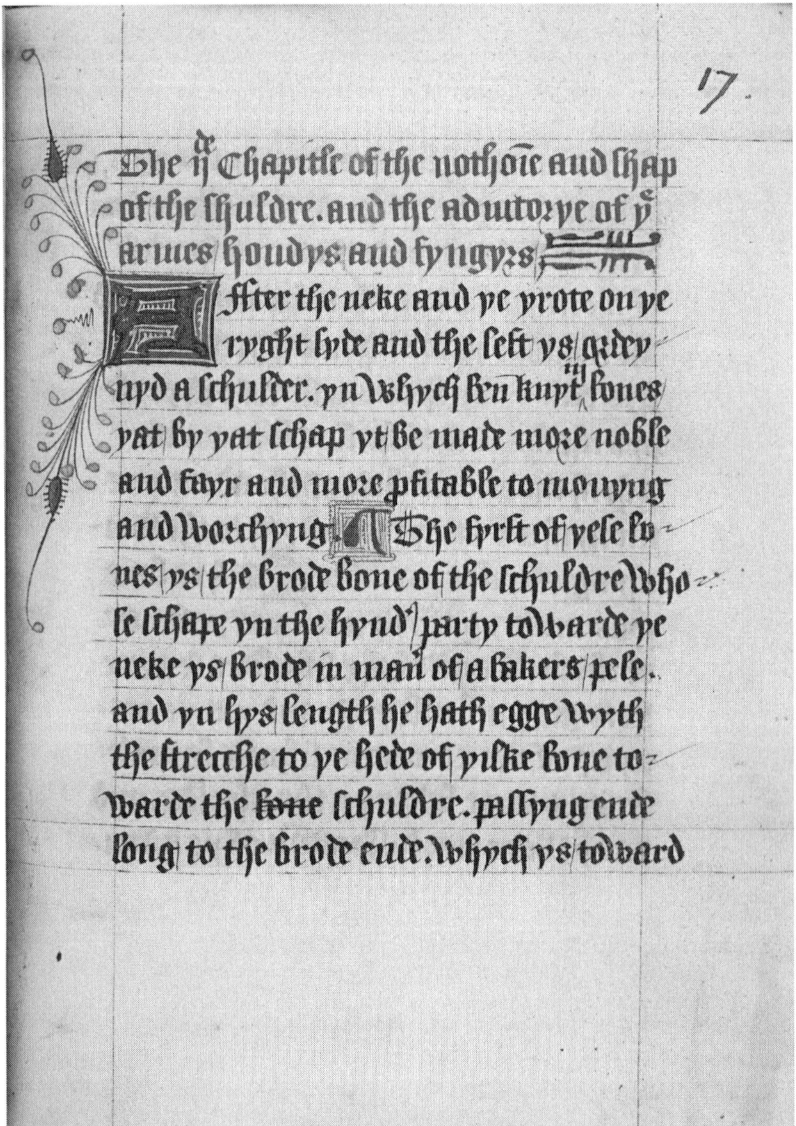

The ij^de Chapitle of the nothome and schap
of the schuldre. and the adiutorye of y^e
armes hondys and fyngyrs
Affter the neke and ye throte on ye
ryght syde and the left ys gaderyd
a schuldre. yn whych ben knyt iiij bones
yat by yat schap yt be made more noble
and fayr and more profitable to mouyng
and worchyng. The fyrst of yese bo-
nes ys the brode bone of the schuldre who-
se schape yn the hynder party toward ye
neke ys brode in maner of a bakers pele.
and yn hys length he hath egge wyth
the strecche to ye hede of yilke bone to-
warde the ~~bone~~ schuldre. passyng ende
long to the brode ende. whych ys toward

Fig. 1.31. Sample folio of W290 (f. 17r)

CHAPTER TWO

THE LANGUAGE

Middle English exhibits "by far the greatest diversity in written language of any period before or since", where "variability was very wide-ranging at every linguistic level: spelling, morphology, syntax and lexicon" (Baugh 1959: 189; McIntosh 1989: 7-8; Milroy 1992: 156; Smith 1999: 121). All this considered, the present chapter has been subsequently divided into three parts: the first section is concerned with inflectional morphology in W290, where the five main word classes will be described; the second section deals with word order; and finally, section three offers a dialectal analysis based on the methodology proposed by the *Linguistic Atlas of Late Middle English* (McIntosh et al. 1986).

2.1. Accidence

2.1.1. Nouns

During the transition from Old to Middle English, the noun inflection had been radically reduced, as case-marking (except for genitive) had vanished and most declension classes had been levelled (Brunner 1970: 45; Lass 1999: 140; Barber *et al.* 2009: 167)[12]. In the Middle English period, a massive reduction of noun paradigm types had taken place[13]. Putting aside minor categories like umlaut and zero plurals, two major declensions took over the noun paradigm (Lass 1992: 111):

	Type A	*Type B*
Sg. Nom. and Acc	-ø	-e
Sg. Genitive	-es	-e
Sg. Dative	-ø	-e
Pural	- (e)s / -is / -ys	-es / -en

Table 2.1. Nominal morphemes in Middle English (Lass 1992: 111)

Table 2.1 above reproduces the general paradigm for nouns in Middle English, which is an inheritance of the Old English period. Type A, on the one hand, is a modified continuation of the old masculine *a-stems*; while type B, on the other, derives from the feminine *ō-stems* and certain weak *n-stems* (Lass 1992: 111). As for the nouns appearing in W290, the singular paradigm selects either for *-ø* or for *-e*.

	Number of instances	*%*
-ø	1048	69,17
-e	467	30,82

Table 2.2. Singular morphemes in W290

The plural, on the contrary, shows a wider range of inflections:

	Number of instances	*%*
-(e)s	366	72,61
-ys	126	25
-is	4	0,79
-n	1	0,19
umlaut	7	1,38

Table 2.3. Plural morphemes in W290

As can be gathered from table 2.2 above, zero inflection more than doubles *-e*, with 69,17% and 30,82% of occurrences, respectively. Concerning the plural morphemes, the preferred plural morpheme in W290 is *-(e)s*, amounting to 366 occurrences. The scribe also employed *-ys* quite frequently, although it only represents 25% of plural endings in the text. Inflections like *-is* or *-n*, as well as umlaut plurals, rarely occurred.

2.1.2 Adjectives

There was no distinction between the singular and the plural in the weak declension of adjectives and, in the strong declension, only monosyllabic adjectives ending in a consonant presented distinctive forms. Thus, the plural for strong adjectives had two forms, namely -ø and *-e*, athough they were used interchangeably, with no function at all. Consequently, the singular paradigm of adjectives shows both inflected and uninflected forms,

(1) From the schulder veyne cu*m*yth anoþ*er* **gret** open veyne . whych goth by the opyn places of the adiutorye and of the arme . (f. 21v).

(2) It ys also to be knowne how þat the **commone** veyne seruieth to the vpper perty and to the nether . whych schewyth hym opynly in the middyl in þe bowȝt of the arme þat ys for he growyth of the schuldre veyne aboue . (f. 23v).

The same picture is observed in the plural, where utterances such as (3) and (4) below are not rare at all in W290, even though the adjective *open* appears arbitrarily inflected and uninflected, showing no agreement with the modified noun:

(3) For thys fyrst uentrycle ys deuydyd ynto ij ***opene*** pertys . (f. 1v).

(4) After the synews and muskles yn the neke þer bene ordeyned veynes and arteryes **open** and preuy veynes and arteryes . (f. 14r).

From a quantitative point of view, uninflected adjectives are the preferred form, both in the singular and in the plural paradigm:

Strong/Weak forms	*- ø*	*-e*
Singular	198 (62,65%)	118 (37,34%)
Plural	52 (68,42 %)	24 (31,57%)

Table 2.4. Adjectival morphemes in W290

392 adjectives have been found in W290, and the two possible morphemes are represented in the table above[14]. As shown, the zero inflection doubles *-e* both in the singular and in the plural paradigm (62,65% and 68,42% over 37,34% and 31,57%, respectively). This is due to the fact that W290 was written in the middle of the 15th century and, by the end of the Middle English period, the adjective had become an uninflected word (Baugh 1959: 192).

2.1.2.1. Comparative and superlative

Comparatives and superlatives in W290 are mostly formed, as today, by the addition of *–er*, previously *–(e)re*, and *–est* to the positive form. Where the vowel is long, it is often shortened in the comparative and superlative, with the final consonant doubled when the stem vowel is long (Burrow and Turville-Petre 1996: 30):

Positive	gret
Comparative	gretter
Superlative	grettest

Table 2.5. Degrees of comparison in W290

Furthermore, there are a two adjectives which change their stem vowel when forming the comparative:

Positive	strong / long / hard
Comparative	strenger / lenger / herder

Table 2.6. Umlauted comparatives in W290 (Lass 1992: 116)[15]

Apart from these forms, a few comparatives are formed with the periphrastic form, i. e. *more noble* / *more profitable*. In addition, there is one instance in which the scribe employs the periphrastic form together with the ending for the comparative (*more strenger*).

2.1.3. Personal pronouns

The personal pronoun is the only word class that consistently, throughout the history of the language, has maintained inflection not only for number and genitive/non-genitive, but also for other cases and gender as well (Lass 1992: 116). However, the inflection of pronouns suffered a great simplification in Middle English. The following table reproduces the personal pronoun paradigm in W290. The forms of the third person present a higher number of occurrences, while the others are just sporadically used.

	Nominative	***Objective***
1st sg.	I	me
2nd sg.	thou/thow/þou	-
3rd sg.	he	him/hym
	she	-
	it/yt/hyt	it/yt/hyt
1st pl.	we	-
2nd pl.	-	-
3rd pl.	thei/þei/þey	hem/them/þem

Table 2.7. Personal pronouns in W290[16]

As can be observed above, several forms are encountered in the 2nd person singular, the 3rd person singular and the 3rd person plural. In order to determine the most frequent form for those persons, the number of occurrences of these items in W290 has been calculated in table 2.8 below:

	Occurrences (%)		
2nd person sg. Nom	thou (11,11)	thow (33,33)	þou (55,55)
3rd person sg. masc. Obj.	him (13,04)	hym (86,95)	-
3rd person sg. neut. Nom/Obj.	it (2,98)	yt (91,04)	hyt (5,97)
3rd person pl. Nom.	thei (2,56)	þei (92,37)	þey (5,12)
3rd person pl. Obj	hem (66,66)	them (18,18)	þem (15,15)

Table 2.8. Personal pronouns in W290

The forms of the 3rd person plural pronoun can be of three types, according to Mossé (1952: 58): 1) *The native type* (typical in the South, it uses the forms of Old English); 2) *The Scandinavian type* (typical in the North, taking Scandinavian *th-* forms); and 3) *The mixed type* (typical in the Midlands, where *th-* forms are used for nominative and *h-* forms for the oblique case). The forms of the 3rd person plural pronoun in W290 present a mixed type, where only the objective form *hem* comes from Old English, the *þ-* and the *th-* forms being Scandinavian.

Table 2.8 above shows that there is one form for each personal pronoun that clearly stands out. Thus, in the 3rd person singular, *yt* predominates in the neuter nominative and objective, while *hym* stands out in the masculine objective. In the 3rd person plural nominative, in turn, the Scandinavian form *þei* is preferred, while in the 3rd person plural objective the Old English form *hem* is still dominant.

2.1.4. Adverbs

The most common adverbial ending is *-ly* (x47), as in *sothly* (x5), *specyally* (x6) or *opynly* (x17), among others. The suffix *-lych(e)*, however, is seldom witnessed in W290, amounting up to just two instances, *prencypallych* (x1) and *properlyche* (x1), respectively.

2.1.5. Verbs

Verbs in Middle English, as well as the other word categories, also underwent a complex process of simplification which,

> shows a characteristic English (and to some extent Germanic - except for German and Icelandic) tendency: a move away from the multiparameter inflection typical of the older Indo-European languages to a restricted

system with one exclusive or dominant parameter per part of speech (Lass 1992: 123)

2.1.5.1. Finite forms

Finite forms in Middle English follow the tendency of the period, which is marked by this levelling of inflections and the weakening of endings (Baugh 1959: 194). W290, being a descriptive treatise, contains exclusively finite forms in the present tense, with only one instance in the past tense (*hadde*). As for the paradigm of the present indicative, several morphemes are found in W290, which are shown in Table 2.8 below.

	Singular	*Plural*
1st person	-e	-en, -e, -yn, -yth
2nd person	-st, -t	
3rd person	-yth, -th, -e	

Table 2.9. Present indicative morphemes in W290

Table 2.9 above shows the different morphemes for the conjugation of the present indicative in W290. The first person singular features *-e*. The second person singular, in turn, adopts *-st* or *-t* (*mayst*), while the third person singular shows a more heterogeneous picture, as it can make use of three different morphemes for the singular (*sendyth*, *moueth* and *clene*) and four for the plural (*comen*, *comene*, *cummyn* and *goyth*). Even though the paradigm for the third person singular seems inconsistent *a priori*, quantitative analysis demonstrates that there are clear preferences for some morphemes.

	Occurrences			
3rd person sg.	-yth (51,14)	-th (37,44)	-e (11,41)	
3rd person pl.	-en (49,18)	-e (39,34)	-yn (4,91)	-th (6,55)

Table 2.10. The present indicative morphemes in W290.

Table 2.10 shows that *-yth* (51,4) is the preferred form for the singular and *-en* (49,18) stands out in the plural. Endings such as *-e* for the singular and *-th* for the plural are seldom found in the witness, however.

2.1.5.2 Non-finite forms

When it comes to the inflection of the infinitival form of the verb, the endings *-e* (x19) and *-ø* (x17) happen to be the preferred forms, as in *wette* and *bring*, respectively. In addition, the infinitive may also be formed with *-en* (x8), as in *fonden*.

The present participle is formed with *-yng*, as in *takyng*, *hauyng* or *coueryng*, among others. The Anglo-Saxon *-end* is not witnessed in W290, as it was progressively replaced in ME by *-ing*, first in the South and Midlands and later in the North, owing to the influence of *-ing* verbal nouns (Roura 1999: 207; Calle-Martín and Miranda-García 2012: 57).

Past participles present a much more heterogeneous paradigm inasmuch as 6 different forms are found, together with irregular verbs. Table 9 below reproduces the occurrences and the percentages of the different morphemes and the irregular forms of past participles in W290.[17]

		Occurrences	%
Weak verbs	***-yd***	155	33,47
	-ed	119	25,70
	-id	3	0,64
Strong verbs	***-en***	23	4,96
	Ø	12	2,59
	i-	6	1,29

Table 2.11. Past participles in W290

As can be drawn from table 2.10, past participles in *-yd* (155) stand out in W290, as in *clepyd* and *endyd*. Past participles in *-ed* (119), in turn, also occur quite often, followed by *-en* and uninflected forms, with 23 and 12 occurrences respectively.

W290 also features the participial prefix *-i* (*iturned* or *itaken*, among others), although its low frequency demonstrates that it was in a process of obliteration. Finally, irregular verbs are very frequent in W290 insofar as they constitute almost half of the past participles in the manuscript (*done*, *gone* or *kept*, among others).

2.2. Word order

From a formal viewpoint, Middle English could be defined as a transitional stage in the history of English, right between "English as a typologically 'Old Germanic' language and English of the type now familiar to us" (Lass 1992: 23). Thus, while "in its early stages English was a heavily inflected language with a relative free word order and a lexical base of mainly Germanic words, nowadays it has become a language with a morphology and syntax radically different" (Fischer and van der Wurff 2006: 109). Due to the rapid loss of inflections, this drastic change "started in the late Old English period in the northern part of the country and was more or less concluded in the fourteenth century" (Fischer 1992: 207).

2.2.1. The noun phrase structure

The organization of the different constituents within the clause is not very distinct from Present Day English, where a noun phrase may contain a predeterminer, a determiner, a postdeterminer and modifier together with the head of the phrase. The following table, partially adapted from Fischer (1992: 211), accounts for the noun phrase structure in W290:

Predeterminer	***Determiner***	***Postdeterminer***	***Modifier***	***Head***
all	þe			veynes
	a		soft	cloyth
	the	xxij	hynder	spondels

Table 2.12. Noun phrase structure in W290

In spite of this general tendency towards the above organization, W290 also shows some instances in whihc the head´s modifier comes after the modified head, as in *the uttermore uertues and instrument sencyble* (f. 2r) or *yt ys bonden wyth strong bondys vnsencyble* (f. 38r).

2.3. Dialect

The present section will analyse the diatopic plane of the language in W290, that is, the variation of language across space[18]. From a historical viewpoint, three major approaches have been made to Middle English dialectology.

Oakden (1930) covered a period of 300 years (1180-1486), although he concentrated mainly on the 14th century. He studied forty-five items on the basis of literary texts, a few local documents and place names. This study presented two shortcomings: a) diatopic and diachronic variation are confused due to such a wide chronological span; and b) the original material is not made available for the reader so that the verification of some data results impossible (Fisiak 1983: 200; McIntosh 1989b: 22-27).

Some years later, Moore, Meech and Whitehall (1935) used 309 literary texts and documents covering the period 1200-1460, focusing mainly on the 15th century[19]. They studied just eleven items so that it could be said that this project and that by Oakden shared the same problems: a small number of items and a wide chronological spread (Fisiak 1983: 200; McIntosh 1989b: 22-27).

The *Linguistic Atlas of Late Middle English*, *LALME* for short, (McIntosh et al. 1986) covers the period 1350-1450 and it is exclusively oriented towards written English. Out of a total of 1150 MSS, the number of studied items amounts up to 270. As for the typology of texts, both literary texts and local documents have been included (Fisiak 1983: 201; McIntosh 1989b: 22-27). Therefore, when it comes to English historical dialectology, and Middle English in particular, *LALME* is now deemed the masterpiece[20] to ascertain the likely provenance of a given text. As Samuels describes it,

> The largest single contribution of this survey [*LALME*] to the history of English is that it provides us with a frame of reference for isolating and classifying those types of language that are less obviously dialectal, and can thus cast light on the probable sources of the written standard English that appears in the fifteenth century (1989: 66).

From a methodological point of view, once the transcription of the witness was completed, the next stage was the tagging and lemmatization of the corpus. Finally, the tool *Text Search Engine* (Miranda-García and Garrido-Garrido 2013) is used to automatically generate a lemma-based list of allographs, which are later taken as the input for the *LALME* questionnaires, according to the 'fit-technique'. First, the questionnaire is filled in by taking some of the words in the text, mainly functional words; secondly, these items are placed in the dot maps, thus providing a likely area of composition for that text; next, the results are checked against the item maps, which afford a more specific localisation of the variants of a word. Finally, the county dictionary confirms a likely provenance for the text in the light of other items which have not been previously dotted in the item maps (Benskin 1991: 24).

In order to generate the linguistic profile of W290, items from the general section of *LALME* were taken and, since their spelling pointed to the east-southern area of the country, items from the southern section of *LALME* were added to this questionnaire. In table 2.13 below, part of the linguistic profile of W290 is partially displayed, providing the allomorphs of each word with their corresponding number of appearances in brackets.

AFTER	after (52) | Affter (6)
ANY	eni (1)
BUT	but (23) | butt (1)
EACH	eche (1)
FLESH	flech (6) | flesch (2) | flessh (1)
FROM	from (55) | fro (16) | frome (3)
I	I (1)
LESS	lesse (6) | lasse (1)
MAN	man (6) | man*e* (2) | mannys (1)
MUCH	much (2) | moche (1)
NEITHER	neyþ*er* (1)
OTHER	other (13) | oþ*er* (8) | oder (1)
SHALL	scholde (4) | schuld (4)
THAN	þan (6) | than (1) | þanne (1) | thanne (1)
THEM	þem (4) | them (3) | þ*em* (1)
THEN	þen (1) | þen*e* (1) | thenes (1)
THEY	þei (36) | þey (2) | they (1)
UNTIL	tyl (1)
WELL	wele (5) | wel (2) | well (1) | welle (1) | weyl (1)
WHEN	whan (20) | whan*e* (1) | whanne (1)
-LY	aliuelych (1) | clothelych (2) | dedelych (1) | p*re*ncypallych (1)

Table 2.13. Linguistic profile of W290 (partial)

The items were selected according to two basic principles: the degree to which they display regional variation, and their frequency, so as to work with a questionnaire which may offer enough discrimination to localise the area in which W290 was written (McIntosh et al. 1986: 1-50).

Consequently, 20 items were selected on the basis of the above-mentioned criteria. The linguistic profile was limited to the following series of allomorphs: *after* (after), *any* (eni), *but* (but), *each* (eche), *flesh* (flech), *from* (from), *I* (I), *less* (lesse), *man* (man), *much* (much), *neither* (neyþer), *other* (other, oþer, oder), *shall* (scholde, schuld), *than* (þan, than, þanne, thanne), *them* (þem, them, þem), *then* (þen, þene, thenes), *they* (þei, þey, they), *until* (tyl), *well* (wele, wel, well, welle, weyl), *when* (whan, whane, whanne) and adverbs in *-ly* (-lych). In the light of these 20 words, the data obtained led us to conclude that W290 seems to have been written somewhere in the counties of Norfolk, Suffolk, Essex, London or Hertfordshire.

Next, the item maps of *LALME* were employed with the intention of narrowing that area into something more specific. Thus, some other relevant words were plotted in zone 6 of the item maps. The spelling of the words in Table 2.14 below leads us to the conclusion that W290 is likely to have been written somewhere to the north of river Thames in the county of Essex[21], nearby the city of Chelmsford, thus following the conventions of the East Midlands dialect.

BE	are (2), ar (1)
GOOD	gode (1) | goode (1)
IF	yf (4)
LITTLE	lytyl (10)
MANY	many (4)
MIGHT	myght (2)
NOT	not (33)
SHALL	schal (6) | schalt (4)
THESE	þese (41)
WH-	whych (158) | whan (20) | what (4) | where (2) | wherfore (4) | why (3)

Table 2.14. Inventory of words plotted in zone 6 of the item maps (LALME).

CHAPTER THREE

THE TEXT

3.1. Editorial conventions

The present section provides a detailed list of the conventions followed in the edition of W290. The principles of a semi-diplomatic edition have been followed so as to offer an accurate reproduction of the original. Editorial intervention has been kept to a minimum so that it can be used not only for research purposes in Linguistics, but also in other fields such as Ecdotics or the History of Medicine.

A graphemic approach has been used, thus employing one single symbol for each letter, even though the scribe writes down different allographs for the same letter, as in the case of <ſ> and <σ> for <s>, or the different allographs for the letter <r>. The guidelines below, partially adapted from Petti (1977: 24-35), have been followed:

a) The spelling, capitalization and word division have been retained, including the use of <u> for <v> and <i> for <j>. The different spellings of a same consonant, however, have been regularized, as in the cases of <s> or <r>.

b) Punctuation has been also kept as in the original witness, where the period and the paragraph mark have been found, the latter being normalized with the use of the symbol ¶.

c) Abbreviations have been expanded with the supplied letters italicized, the superior letters have been lowered to the line, as in þ^{t} > þ*at*.

d) Emendations, which consist in the cancellation of single letters as well as full words, as in ku~~y~~ttyng (92) or ~~elepyd~~ (142), have been reproduced in the apparatus (*del.*).

e) Scribal errors are also represented in the apparatus (*dup.*). They consist in repetitions of words or phrases, as in ‘and’ (69) or ‘And that the rubbyng’ (921), respectively.

f) Insertions, which are above the line in the original, have been lowered to the line and placed in their corresponding position within the text, which is marked with the caret (^) by the scribe. Thus, those insertions have been marked using slashes, as in ‘fa/s\tened’.

g) Marginalia is included in the critical apparatus, except for the numbering of the folios. The words appearing in the marginalia are either catchwords or Latin words, which were included in the manuscript as a kind of index.

3.2. Constantinus Africanus' Venerabilis Anatomia

hic incipit liber uenerabilem Anotome f.1r um secundu*m* Galielem . Fyrst of the brane and the hede and membres beyng aboute them or yn*e* them .

Sothly þe brayn ys soyft yn hys substance and marowhy hauyng long schape after the lengthe of the hede . And yt ys all dep*er*ted yn to thre partyes . yn to the forpart medyll party . and the hynder party . whych p*er*tyes ben*e* clepyd uentrycles or lytyl wombys wyth som rowndnesse . And þese uent*ri*cles ben*e* holownesse yn whych ben*e* the spyrytys of the sowle . And yn wych þei resauyn alterac*i*on and dygestyon . vpon that þat þei resauyn*e* yn the hert to þat f. 1v that þei ben*e* able to fulfyl the worchyng*s* of þe sowle . And all þougth þes uentricles yn comon*e* be nowmbred iij . neuerthelese the formor party whych ys more þan þe other ys opynly deuydyd ynto ij . part*s* . wherfore þese uentryclys of the brayn ben*e* seyd to be iiij . And al thowgh þe oþer ij wentryclys haue departyng*s* þei be not seyd to haue dyuysyons . For þ*i* þat þey ben*e* hedde and ful on*e* sencyble . but yn þe dep*er*tyng of þe fyrst uentricle yn to þe oþ*er* ij . be not so . For thys fyrst uentrycle ys deuydyd yn to ij open*e* p*er*tys . ¶ In the fyrst p*er*ty of the fyrst uentrycle of the brayn ys ordenyd the vertu þat ys clepyd co*m*e wyt or fantesye . whych taken all the fourmes taken the . v . wytt*es* f. 2r and demyth of hem aft*er* þe remouyng or p*er*sens of the utt*er*more uertues *and* instrument*s* sencyble . ¶ In þe secunde p*er*ty of the fyrst ventrycle ys ordenyd ymagynac*i*on that holdyth to gydder or kepyth the fourmes Itaken*e* of comon*e* wytt . In the ij wentrycle or yn the medyll is ordenyd thought estymac*i*on ordeinyng . ¶ In the last or yn þe iijde . ys ordenyd mynde . Anone vpon*e* the brayn ys ordenyd a pannycle or a soft cloyth and neu*er*thelasse harder in substance thanne ys the brayn . that by the fence or sekernys of þis pannicle the hardenesse of þe f. 2v vpper pannicle and the bone ne greue nor defende the brayn . And þis pannicle[1] or softe cloyth ys wouyn togyddyr yn man*er* of a nette wyth wanys and artyres byndy*n*g and holdyng ij arteryes and vanis fast to gyddyr . by godenys of hys makyng clothelyk . And he ys dep*er*tyd from the brayn . and yn sumplace contynued wyth the brayn by vaynys and arteryes . wych go oute of þat pannicle . and entren diuysyons of the brayn*e* . and quyken*e* and inbylen her substance . and þeuen to hem aliuelych spryte . whych ys dep*er*tyd and alt*er*atyd from the kynde of the brayn . after that hyt cordyth to vertues of worchy*n*g of the soule . And þis forseyd pannycle or f. 3r ful softe cloyth nexte þe brayn ys clepyd pia mater . Or the myryng vpon whych pia mat*er*[2] . ys a veile or a nother pannicle[3] wouen also to geder in man*er* of a nette of veynes and[4] arteryes fast to gydder byndyng and holdyng by godenesse of hys clothelych makyng . whych ys clepyd dura mater . ys harder than ys þe pia mat*er* . so þat yt defendeth the brayn*e* . and the pia mat*er* from the hardenesse of the bone . that yt take no harme . ne hurtyng of thylke bone . but be defendyd by hys harde-

[1] **50** pannicle] *out. marg.* pia mater
[2] **67** pia mater] *out. marg.* pia mater
[3] **67** pannicle] *out. marg.* dura mater
[4] **69** and] *dup.* and

nysse . and he ys dep*er*tyd and not co*n*tynued wyth the fyrst pannicle but in sum place . by whych passen subtyle veynes and arteryes to the helpyng of the softer and the brayn . Ne thys vpper or herder pannycle f. 3v ys not conteyned wyth the bone of the brayn panne . lest yt take co*m*mone hurtyng for þe herdnesse of hyt . ¶ But wyth som ioynters medelyng of the skul . by whych the clothy bynedyng . gone oute wyth arterys and smale hery vaynys . And of these p*er*tyes of þe clothy byndyng ys made þe vtt*er*mare skyn . keu*er*yng al the skul as yt chewyth . ¶ Now howe grete doute and drede fallyth yn the kuttyng[5] of skyn vpon the ioyntes yn the meddelyng of the skul bone . and of p*er*syng wyth tre the panne and the raffe yn the placys of the ioyntes . or meddelyng . For why by the kuttyng or p*er*syng yn suche placis f. 4r co*m*meth noysaunce or hurtyng to the dura mater of the brayne . Therfore be þer nor worchyng wyth yren*e* in such places yf yt be possible to be eschuyd . ¶ Ferthermore wete thou wele that[6] the brayn yn the formar p*er*tye vnder þe bone of the forhede hath ij . addicyons lyke hedys of bygg yn the whych þe worchyng and þe vertue of smellyng ys fulfyllyd . And þ*er* aboute the flesshy ende of the nose ys a herde coueryng of the vppermost skyn *and* ys helyd þat þe sup*er*fluyte of the fore partys goyth oute by ij vaynys . And also the same oure cou*er*yng ys made holow yn p*er*ty goyth to the palet . that the sup*er*fluytes of þe f. 4v secundyne and last ventrycles ben pourged by . ¶ Two veynes vppon þis two clo-

[5] **92** kuttyng] *del.* ku~~y~~ttyng
[6] **102** that] *out. marg.* braine

thes . that ys to saye . dura mater . and pia mater ben*e* ordeyned .

The bones of the skull ben*e* . vj . and one yn the ende . whych ys be twe-ne the fyrst bone in the begynnyng of þe chyn . *and* the bone of the hede . whose schape ys lyke to the latter . Alauda . And he ys vnder beryng in the hynd*er* party al the bonys yn the hede . and berer vp of the paxillus . And the fyrst bone of the skul of the forhede ys clepyd coronale . and yt ys schaape yn man*er* of an halfe cercle . And hath sum goyng oute yn the for- f. 5r party . *and* he ys fa/s\tened wyth the bone of the nose yn the forparty in the maner of a sawe . *and* þ*er* ys mayd a schap of faste-nyng in the man*er* of a cros . whose vpper-more semyth long .

The ij . vernaylle bones ben faste-ned in the myddys in the hede in man*er* of a sawe be twene hem selfe w*ith* the bone clepyd . alauda . and vnder þis bone ys put the boyn*e* þat ys clepyd paxillus . Ne ys not fastened wyth hym . but vnderberyth hym and ful-fyllyth the schap of the hede . ¶ This bone paxillus . ys[7] aplyed to the bone cleped . Alauda . and ys put and fastened be twyxte hem and the f. 5v ouer chanel . And by thys man*er* of faste-nyng and puttyng to gedder of these . vnd*er* whych be the erys . and þei ben cle-pyd ossa mendosa . whych be not faste-nede wyth the werneyllys and gret bones . but by puttyng on of party to party . ¶ And be nethe þese bones a-nenste the vpper chawel ys a grete ho-le . whych ys clepyd peterosum . and yt ys to sey ful herde and stony . whych ys the substance of the bone . mendosa .

[7] **142** ys] *del.* clepyd

¶ By whych passyth the synewe of heryng . here ben therfore nowmbred . vj bonys comp*re*hendyng the brayne . And one that ys applyed þ*er*to clepyd . paxillare . ys yn the hynder partie beryng vp the other vj . and fulfyllyng f. 6r the schape of the hede . ¶ Anone nexte vpon these bonys ys made a skynne of pannycles wythynneforth . and fastened wyth the synues . And þat pannycle or skyne ys made thynne . and spred apon al the bonys . and byndyth hem from wythoute forthe . ¶ Apon this skynne ys made a thykke heryd skyn . and yt ys wyth venyes and arteryes . and synuysse . co*m*myng from wytheoute forth . and specyally yt ys wonden of the synue co*m*myng from the hole of the fyrst spondyl . and of the secu*n*de . whych ben medelyd wyth the veynes arteryes . and skynnys . wenyng to gydder the skynne . kou*er*yng the skul . and f. 6v thys skynne ys made harde . and ys hauyng hym selfe thynne . lytyl flech . not wele sygthty . whych makyth þe gretter skynne and membre . ¶ Wete þou wel þat al the kuttyngs of the hede skul be lettyr . and of better schape þat be mayd aft*er* the here renneth . For why . the senues of the flessh skynne lyen that wey for the moyst p*er*ty .

Also yt ys to know eu*er* þat the kuttyng whych schuld be mayde hyt ys not nedeful to be in eu*er*y me*m*bre after the graynes . as yn the forhede . For why . in the forhede although the greynes gon*e* ouerthwert the kuttyngs . And namely depe owen to be done after f. 7r the contrary of the greynes . for the synues of the forhede . gon*e* after the length*e* . and not after the greynes . ¶ In the forp*er*ty of the hede ys ordeyned the nose . whych ys made of bone and grystle . and faste-ned wyth the holes . wyth the ij . schew-yng oute of the senues in man*er* of hedys of bygg*er* . In the whych the worchyng ver-tue of smellyng ys fulfyllyd . ¶ And þese synuesse accordyng be dep*er*tyd of þe iijde . payre of synues . þat ys wyth yn þe nose . and aboue . acordyng to the felyng place . ¶ wyth oute thys nose yn the palet hath an hole . and the holys of the nose ben*e* contunyed wyth the same hole . f. 7v thatt whanne the mouthe ys shet . the eyre may passe by the euentyng of the herte . *and* also that yn tyme of sclepe man may bre-the . ¶ There beyth ordeyned vnd*er* the bo-ne of the forhede in the forp*er*ty . ij . þen*e* . whych be mayd and ordeynd of iij humours . and vij . cotys in thys man*er* .

Off the forme/r\ brayn ben*e* departyd ij . holough synnues . and þei ben of the fyrst payre of synues . hauyng her waxyng of the brayne . and ioynen þ*er* to gydder . and maken one holownesse . After þei ben dep*er*tyd in here outegoyng from þe skul . and þei beth cloþed of pannicles of þe brayne . ¶ And whan þei gon*e* oute of the skul þ*er* ys made of hem a thykker and f. 8r a harder pannicle . whych ys clepyd sky-rotus . ¶ After thys ther ys made of þis anoder pannicle . whych ys clepyd secun-dina . And thys skynne ys clepyd so forth that he ys sett and made after the fyrst skyn . and he comp*re*hendyth wythinne hy*m* the glasyn humors of the Iþeu*m* . And after thys skyn ys made another skyn whych ys clepyd . Retinus . forþi þat he ys made of such a schape . ¶ And þis retinus comp*re*hen-dyth ym hym halfe the crystallyne humo-re . and of thys skynne retinus ys congen-dred and made a skyn*e* clepyd araneus . and

yt ys clepyd so forthi . þat that skyn hath
such a schape . and he co*mpre*hendyth wythyn
hym þe oþ*er* halfe of the crystallyne humore .
After retinus þ*er* ys another skyn*e* f. 8v
whych ys clepyd . vneus . forthy
that yn hys schape and dysposyc*i*on yt se-
myth to haue a lyknes wyth . vnea . And
thys ys holowe yn the myddys . And þis
hole ys clepyd pupilla . and yt closeth
and oponeth aft*er* yt nedyth þe þ*er*to . þat the
syght be fulfyllyd in the Crystallyne humo-
re . and yt comp*re*hendyth in hym selfe all
the humoure . ¶ Albugeni*us* . whych ys
nedeful to the kepyng and defence of the
Crystallyne humoure . And thys humo*ur*
Albugeni*us* . schuld go oute of the hollow-
nesse of vnea . but yf yt were couered *and*
so of nede ys made engenderyng of a noþ*er*
skyne þat ys corneus . whych ys clepyd[8]
so for lyknesse þat he hath to a whyte horne . f. 9r
¶ Thys skyn ys made of the skyn skly-
roticus . and yt ys wyth him byndyng
all thyþe . and for thys byndyng was not
wel sykyr . ¶ Kynde wolde for the best þat
þ*er* ware streng*er* byndyng or skyn made .
whych schuld be strenger þan all . and so
ys þ*er* a skynne made . whych ys streng*er* and
all the oþ*er* . and ys clepyd . Coniu*n*ctiuus .
And þat co*mpre*hendyth all the Iye . saue þe
blake . and yt ys made of the skynne co-
u*er*yng the skul .
FRom wythoute forth we tolde
þer was made of the p*er*tyes of
the skynne . co*m*myng be the medelyngys
and Ioyntes of the skul . fro the skynnys
of the brayne . And so semeþe by þis maner f. 9v
þat yn the re*m*meuyng of humors . from the
and the brayn . hyt helpyth to knyt the
vanyes of the forhede . ou*er*thwerte . ¶ Aft*er*
thys þ*er* co*m*myth the Iye synuesse of þem

[8] **256** *bot. marg.* so for

two payre senues of the brayn be the holes of the skul of the Iye . whych þeue the mouyng to the Iyen and syght . By whyche þei se mouyng thyng*s* . Thow mayst also wete by thys that blode lettyng of the veyne of the forhede . and also blodeletyng of þe veynys of the tery places of the eyn þat ben*e* be the nose helpyth to sekenesse of þe Iþhen of the Iye kynde . And the re*n*nyng of hem to places of the Iyen . ¶ Sothly f. 10r vnder the Iyen in the forparty and vnd*er* the nose ys the chanel . whych ys made of . xiiij . bones . of whych the Ionyng and makyng ys hydde . and not wele knowen*e* . wherfore of þat makyng and dep*er*tyng of hem . we make no mynde but yn honde worchyng . þei be put vpon as oo bone . In the vpper chanel ben fast xvj . and yn som xiiij . vpon the bone or bones of the vpper chanel ys made a skynne of the skyn kou*er*yng the bone of the forhede . and of iij . payre synewes of the brayn . and of p*er*ty of the iiijth payre . and the p*er*ty nexte tyll hym . ¶ And the ijde payr synewes of . nucha . w*ith* the iijde . ys made a skynne w*ith* weynes and arteryes couenably coueryng f. 10v hym . and þese senues þat in þis maky*ng* co*m*myng fro the partys of the synewes of the brayn of . Nucha . þeuen felyng and mouyng to the face . and to the palet . and to the nose . *and* moyst whan þese synews entren the makyng of the muscles mouy*n*g þese p*er*tyes . ¶ Therfore yt accordyth . þat the kuttyng þat be don*e* or owen to be done yn the place of the vpper chanel . *and* of the nose . owen to be done aft*er* the granes of þe place . why in thys membre the greynes goon*e* after the goyng of the synews and of the muscles of the face and of the nose . from the ijde . the iijde . and the iiijth . payre senews of . Nucha . co*m*men senews to þe muscles of the Iawes whych mouen þe nether chanel and the cheke bone . ¶ Thys f. 11r ne*þer* chanel ys mayd of ij bonys . whych be yoned to gyddyr in man*er* of a sawe . and yn the hynd*er* p*er*ty . þei ben*e* Ioyned w*ith* the knotty p*er*ty of the vpper chanel . yn whych ne*þer* chanel ben pyngth*e* in som xvj . *and* yn some xiiij tethe . And the vpper tethe ben bonden wyth the Iawys medlyng the byndy*ng* and skynnys co*m*myng frome the skynnys cou*er*yng the bonys of the vpper Iawe . and byndyng hem to gydd*er* . and the skynnes byndyng to gydd*er* the nether chanel i*n* the skynne . For why . all þese skynnys and byndyng co*m*men down from the skynne dep*er*tyd . and kou*er*yng the skul . whych ys mayd of the skynne of the brayn . And þ*er* fore yt semyth þat the akyng of the tethe f. 11v hurteth the hede by the myddys . and the brayn bryngyth in desease in all the hed . by twyxte the vpper chanel and the ne*þer* chanel ys set the tong . whych ys nesche flesch and whyte veyny cordy . or stryngy . and synewy . And yt ys takyng or hauy*ng* of vj . payre of synues of the brayn a sencyble synew of spekyng . and of vij payer of synews . a mouyng synew .

Ther ben*e* in her ij veynes whych schewen hem whan she ys leftvp and þei ben*e* oponed for sekenesse of the tonge and apostomys þat ben*e* yn the rotys of her . But yt ys to take hede here þ*at* fleobotomye of the tong schal not be doo in no cause . but þ*er* be to sore gen*er*al blode le- f. 12r tyng or ventusyng in the schulders or clensyng w*ith* medycyne or clysterye for blode letyng of the veynes of the tong . Draueth and multyplyeth the syknesse to the rote . and specyally þet the body be replete or styptyke . and þet som of þo þyngys go befor . of whych we seyden . ¶ After the chanells and vnd*er* the vernayle bones yn the ryght syde and the left syde ben ordeyned þe erys . vpon the harde holow bone clepyde peterosum . and yt ys of the substance of the bone clepyd mendosa . that ys hauyng yn hys holownys wryncles in hys passyng or weye . And that man*er* yt goyth f. 12v vnto the synew of the fyrst payer of synues of the brayn . by whych synew ys mayd the heryng . *and* þis synew is holow .

Uppon þis growyth and ys made a brode gristyl in the whych ben*e* felyng synews co*m*myng from the skynne cou*er*yng the skul . And þ*er* be com soft flesshy p*er*tyes ther . and þ*er* be com wryncles of weynes and synews halfe rou*n*de ther . whych the leche cut after theschap whan cuttyng ys nedeful . yn that place .

Under the hede yn that place yn þe hynd*er* party ys . Nucha . vnder whych ben*e* ordeyned vij spondels . whych ben*e* vij Ioyntys of the neke . Of the whych nucha and the spondels gon oute vij f. 13r payre of synews . so þat the fyrst payer goth oute of the fyrst hole of the fyrst spondell . the whych hole ys anenste the begynny*n*g of . nucha . where nucha ys dep*er*tyd from the brayn and spred vpon the skyn of the hede . and þ*er* ys made a p*er*tye of þe muskles .

The ijde payr goth from þe ijde . hole þ*a*t ys betwene the fyrst spondyle and the ijde . and goth into the skyn of the hede . and þeueth hym mouyng and yn som p*er*ty yt ys made w*ith* the muscles of the neke . *and* of the cheke bones . and þeueth them mouyng .

The iijde payre goyth oute of the iijde spondele . and yt ys sprede or partyd to the mouyng of the muscles of the cheke bones yn to mouy*ng* of the brode bone of the schulder . f. 13v The iiijth payr of senues goth oute of the vpper hole of the iiijth spon-del fro þat yt goth to the musches of the bake . and to the muskelys of the for p*er*ty and of the hynd*er* p*er*tye . The vth payre goth oute of the vpper hole of the vth spondele *and* ys spred on the muskeles of the mydryfe . and to the muskles mouyng the hede . and after to the muskles of the schulders bla-dys . The vjth payre goth oute of the vp per hole of the vjth spondyle . The vijth payre goth oute of the vpper hole of the vijth spondyle . and yn her goyng oute þei ben dep*er*ted pw- f. 14r ard and downward vpon the muskles of the neke . the hede . the brest . and the throte . ¶ After the synews and muskles yn the neke þer ben*e* ordeyned veynes and arteryes open and p*re*uy veynes and ar-teryes . ¶ In the nek oponly ben*e* two by hynde the eyru*m* . of whych the kuttyng ys dredeful . for of the unkyndly hete of hem wyth the herte and the brayn follow-yth sownyng and dysseasse i*n* many thy*n*g . and of the greate bledyng aft*er* there cut-tyng . þei wyl not lyghtly be stoppyd . *and* so þ*er* folowth deth of tyme . ¶ Therfor yt se-myth oponly þat all the apostomes of þe nek and all the cuttyng þat owen*e* to be done yn the neke . and of kynde and fettyng f. 14v of membres of the neke . and the growyng of the senues yn that place owen to be don*e* aft*er* the length . and the forsayd veynes and arteryes yn whych ben*e* tyed to gydd*er* . owen to be consyderyd forthy . that in þat vndyr the forseyd gret opon veynes gon vnder þe other veynes . so þat the veynes in that pla-ce mow not be cutt or persyd . but the arte-rye were cutt or persyd of whole p*er*syng shold folow deth .

In the forp*er*ty of the spondyls of þe neke be syde the throte ys sett the pype of the stomake . whych ys clepyd mery or ysophagus . by whom mete and drynke goth to the stomake . and yt ys made of f. 15r strenges veynes and arteryes and synews co*m*myng fro the syxte payre of synews of the brayn . and ys fastned wyth the throte w*ith* the pype of longes clepyd . Trachea arteria . whych ys made of gristles of the vjth payer of synews of the brayn . and yt ys greyny . and hath lytyl hoopes yn the vtter party of the fastnyng wyth the meryngg of the pype of the stomake . whych hath euenhed and softnesse . and on the vpper ende he hath coue*r*yng clepyd Epigloton þat þe tyme of etyng the mete entre not in hem . ne no thyng hurtyng but only eyre . or ony thy*n*g lyke to eyre . ¶ And þou schalt vnderstond by epygloton the knote yn the þrote yn the ende of the pype . the whych knotte f. 15v ys schape . þat he hath fallyng down*e and* makyth dyuers voices . ¶ Aboue þe coue*r*yng *and* the pype . þ*er* is made aboue þe rote of the tong a maner i*n*strument þ*a*t clenyth to þe palet i*n* þe ouermoyst p*er*ty of the mouth clepid . vnula . whych is necessarie to brekyng of þe eyre . *and* to brekyng for diu*er*se voyces *and* most whan he holdeth hys kyndely schap fourm *and* qua*n*tyte . and for þ*a*t whan he waxeth ou*er*e gret *and* lettyth the voice . he is kut . þ*a*t he is þe better . instrument to bryng forth diu*er*se voices . vppon theise pipes ben*e* ordeynyd seneuys g*r*ete *and* smale opon*e and* pr*e*uye . whych cu*m*en from the syrte *and* vijth paire synews of þe brayn . *and* ben*e* meddelyd w*ith* þe muscles of þe nek and of the throte w*ith* þe synews of the iijde . *and* iiijth paire sy f. 16r news of . nucha . ¶ On þe ryght halfe *and* the left halfe of þe pipe of þe long . ben*e* ij open veynnys grete *and* blake whych ben*e* clepid gindet . and vnd*er* hem ij arteries of whose cuttyng or p*er*syng þe blode rennyth from þe lung disesyd . for his kynde is lettyd i*n* worchyng and hys hert is desesyd . And by þis cause co*m*meth sodeyn deth . and þ*er*for al þe kuttyngs þat ben mayd yn þe throte for what cause eu*er* yt be . it owe to be don*e* aft*er* þe length and to eschew all þe veynes of þe throte . þi power *and* specyally grete . and þise gyndet . for eu*er*y veyn of þe throte hath vnd*er* hym an arterye hyd . for þe whych althermost *and* nexte he hath kyndely hede w*ith* þe lung *and* the hert . and so eu*er*y kuttyng of þis place ys dredeful as semyth opynly to wete .

wete þou well þat þe throte ys fastnede w*ith* þe forcle of the brest in place clepyd pixis gule . þat f. 16v ys to sey þe box of þe holownese of the throte . and þe neke ys fastenyd w*ith* the hynd*er* partie w*ith* the viijth spondele . whych ys the fyrst spondel of the brest . And also he ys fastned w*ith* the spawd bon*e* vnto the schulder that yt maye moue the bett*er* . and more myghtyly whan yt ys nede and that hys schape be feyrer *and* better .

Dura mater . Pia mater . The neþ*er*myryng . Paxillus . Coronale . Paxillare . Alauda . Ossa mendosa . Peterosu*m* . Retinus . Araneus . Vneus . Vnea . Pupilla . Albugeni*us* . Corne*us* . Schrotic*us* . Co*n*iu*n*ctin*us* . Tery places . id est loca lac*r*inarum[9] .

The ijde Chapitle of the notho*m*ie and schap f. 17r of the shuldre . and the adiutorye of þe armes hondys and fyngyrs .

Affter the neke and þe throte on þe ryght syde and the left ys ordeynyd a schulder . yn whych ben*e* knyt /iij\ bones þat by þat schap yt be made more noble and fayr and more p*ro*fitable to mouyng and worchyng . ¶ The fyrst of þese bones ys the brode bone of the schuldre whose schape yn the hynd*er* party towarde þe neke ys brode in man*er* of a bakers pele .

[9] **510** lacrinarum] *bot. marg.* The secu*n*de

and yn hys length he hath egge wyth the strecche to þe hede of þilke bone towarde the[10] schuldre . passyng ende long to the brode ende . whych ys toward f. 17v the nek in whych brode ende ys knytte a gristyl . and þat grystyl spredyth be syde þe spondyls of the brest . vnto the seuyn spondels of the neke . And on the other ende toward the schuldre þe bone waxeth grete . and i*n* hys ende ys an holownesse whych ys clepyd the box of the schuldre bone yn the whych the rownde hede of the adiutorye turnyth aft*er* þat yt accordyth to the worchyng of þilke membre . ¶ Thys bone was made yn thylk man*er* þat þe bones of the brest and of the schuldre be more strenger yn her beyng . and þat þe mouyng of the adiutorye schulde be made better *and* more syker yn the schulder . and þat þ*er* ware not dislocac*i*on of þe adyu f. 18r torye . for eu*er*y lytyl cause . hyt semyth opynly by þat . that the bone of the adiutorye . schul not be dyslocate bakward . In the forparty of the scholdre ys ordenyd an ende of the forcle . and yt ys tyde wyth the brode bone . *and* he ys the secunde bone of the thre bonys of the schuldre . þat þilk same membre dwellyth the better yn hys strength . *and* beryth vp that place . þat he go not forwarde of lyght cause . ¶ And þis bone ys lesse þanne the broode bone of the schuldre in hys knyttyng . that the schape of þat place scholde be þe farer . and yt schuld no lette the mouyng of the whyrlbone . and so yt semyth f. 18v opynly how by the beyng lesse of thys bone yn that party the rownde of the adiutorye may be dyslocat forward . Affter þese bonys þ*er* be byndyng

---

[10] **525** the] *del.* bone

not fulweyl knowen þat bynden and knyttyn þese bonys to gydd*er* the whyrlbone wyth the box of the breking strechyng . of whych co*n*tynuel departyng the restoryng of the dyslocac*i*on ys lettyd in so much other whyle þat he may not be brought to hys beyng . and he ne wyl lepe aȝen after hys restoryng . ¶ The iijde bone of the schuldre ys the adiutorye . wych ys rownde wyth oute . and holowe wyth inne . Thys bone ys marowhy for hys gretnesse . and hys þinhede . *and* he resayneth moyst*er* hede þat ys [f. 19r] necessayre to him of the marowe . of wych þe vpper ende ys rownde . *and* goth vnto þe box of the schuldre bone . and þ*er* he ys knyt and turneth aboute . beyng by twene hys byndyng wyth other two bonys yn þat place . as yt ys a boue seyd . The other ende of the adiutorye ys fastened w*ith* the Icupyd aboue hauyng a schap lyke to a roll yn þe graues of wych roll the ende of the vpper bone entreth . and vete þou wele þat the nether bone or roll ys lesse strechyng frome the thombe to þe rol of the elbowe . and by neth ys another lenger bone and more grett*er* and strechyth from þe nether p*er*ty that ys to wete from the lytyl fyggyr to [f. 19v] the elbow . and lyeth apon þat other . w*ith* a lytyl beryng out lyke a byrdys byll . in þe length to the adiutorye . and the roll of hyt makyth[11] the schape of the elbowe . whan yt ys bowed . And that was made þat þilk fastnyng w*ith* ende of the adiutorye yn the roll scholde be more streng . and þat yt schold not be dyslocate of lyght cause for supportyng of the vpper focyle . or bone vpon þe hede of the adiutorye .

---

[11] **595** makyth] *dup.* of yt makyth

Thys place ys bonden strongly w*ith* strong bondes and p*re*uy . þat the schape of the membre . and the beyng of the bones scholde be stronge . and þat f. 20r þei mowen*e* abyde whan a man*e* lyftyth bordons lest þat the arme bow not bakward or outward . and þ*er* the neþ*er* focyle or þe more bone aboute hys passyng outtakyth þe schap of the halfe cercle .

Affter the elbow be ordeynyd þe ij focylles whych ben*e* knyt to gydd*er* ende long. the byndyng goyng by twene and the lesse bone goyth ynto the more . *and* be mayd boyth as on*e* . and be ioyned w*ith* þe bones . makyng the wryst whych ben*e* clepyd racett . the whych ben viij . and iiij of them ben*e* ordenyd wythoute the honde to the ende of the ij focylles . ¶ And þe other iiij ben knytte w*ith* the bones of the hollownesse of þe honde and they bounden*e* to gyd- f. 20v dyr . and þese bonys or racettys haue no marow . for the hardenesse and thyknesse of þe substance . but the focilles hauyn marow for the cause tolde in the bone of þe adiutorye . and þough yt schew not in þe lesse focylle in whych ys fonden moystnesse in man*er* of marow . and yt is þer in the stede of marow .

Affter the bonys of the racete of the honde ben*e* ordeyned þe bonys of þe holownesse . whych ben foure . and þei be bondyn nyȝ to gydd*er* w*ith* racett . and þei ben*e* fastned to gydd*er* opynly . knottly . and þat oþ*er* ende w*ith* þe bonys of the thombe . for þe fyrst gret bon of the thome ys knit to the ende of the vpper focyle in the Ioynte . and þat was made þat yt be more strenger i*n* the f. 21r mouynng þan the oþ*er* iiij . and be more semely Ioyned . In eu*er*y fyngyr of þe honde ben*e* iij bones which ben*e* bounden and Ioyned to gydd*er* and knotly . vppon these bonys byndyng ben ordened synews co*m*myng from the vj *and* þe vij *and* viij paire of synews and of spondels of the neke of which wyth lytyl flesch . In þe byng-dyng of þe bones be mayd muscles . Ther ys a grete one and a opyn*e* yn the myddys of the adiutorye . so þat su*m*me p*er*tie ys w*ith*yn . and þe oþ*er* p*er*ty w*ith*oute . of whych þe cordys mouyng þe arme be de-partyd aft*er* diuers partys . And yn the arme ys a noþ*er* open beyng wythoute þe arme . *and* be spredyth þe arme a boute . of whom ben*e* dep*er*ted diuersse stryng mouyng þe fyngurs inwarde and outeward aft*er* þat it nedyth . f. 21v After þese muscles ben*e* ordenyde veynes open and p*re*uy the whych oon ys dep*er*tyd from the veynys of the arme holys and from the veyn þat ys schewyde yn the arme holys . whych passyth by the schuldre in þe nether party . þe syde of the ar-me goth to the vpper p*er*tye of the elbow and yt ys clepid . Cephalica . for he ys of þe vay-ne the whych oo p*er*ty goth vp to the brayn *and* noryshyth the membres of wytt of the brayn . wherfore by thys kynde . yt ys cle-pyd so . From the schulder veyne cu*m*-yth a noþ*er* gret open veyne . whych goth by the opyn places of the adiutorye and of the arme . and by the vtter partye of þe honde . and yt ys clepyd finis brachij . þ*a*t f. 22r ys to sey . the ende of the arme . whych in man*er*[12] afore seyd . and þ*er*for yt ys clepyd so . From the arme hole wyth ynforth cometh a noþ*er* veyne whych goyth by þe Inner syde of the adiutorie vnto the elbow and shewyth hym i*n* þe nether partye of the elbow . and he ys clepyd basilica . *and* goyth by þe nether bone . and þ*er* he ys bran-chyd bytwene the leche and the lytyl fyn-ger . and ys clepyd saluatica . or epatica . or splenatica humerabilis . Cephalica or as-sellata . ¶ By neth grouth and apertyth

---

[12] **673** man*er*] *dup.* in the man*er*

a veyne þat ys yn the bowȝt of the arme þat ys to sey . in the medyl . whych ys clepyd . purpurea . or nigra . or co*m*munis .
f. 22v or sum folk clepyd matrix . And þis veyne *ser*ueth vnto the vpper parties . and to þe nether . and for þat yt ys clepyd so . ¶ vnd*er* þese vaynes so that eu*er*y vayne hyd *and* open þ*er* ys ioyned to hym an a/r*ter*ye p*er*uyly or openly . ¶ Also of þese muscles so y made wi*th* veynes art*er*yes and som flech and sum p*er*tyes of felyng synews cu*m*yng to the makyng of hem þe bettyr skyn . and by openly synews ys made the skyn of þese membres and the bonys be kou*er*ed . And for þat p*er*tie of the felyng synews co*m*myng to the makyng of þem þe better skyn ys mayd felyng . *and* be opynly seen*e* veynes . ¶ Tho open veynes be noryshyd of þese thyng*s* . we haue
f. 23r seyd opynly how yt be houeth the leche to be haue hym selfe in kuttyng of apostummes of þoo places . For why . whan al the muscles synews . lac*er*tys . at*er*ies . and veynes goon after the length from the schulders vnto the endys of the fyngurs . the places and þe apostomes in þese membres askyng kuttyng and cauteryenge after the length . It semyth þat yn the veyne cephalica . whych ys in the bowht of the arme yn the vpper p*er*tye and whych þat schewyth hym in þe vpper focyle . *and* goth forth bytwene þe thombe . and þe nexte fynger *ser*ueth to the hede . and to the partys of yt . and how the veyne þ*a*t ys in þe lauer p*er*tye of the bouȝ*te* of the ar
f. 23v me þat passyth by the nether focille . and yt ys schewyd in the honde bytwene the leche and the lytyl fyngre . and yt *ser*ueth to the lyuor . and to þe splene for þy þ*a*t yt ys dep*er*tyd from a partie of the vayne noryshyng

the[13] lower membres . and yn thys forsayd place of þe honde yt ys clepyd . Saluatilla . ¶ It ys also to be known*e* how þ*a*t the co*m*mone veyne *ser*ueth to the vpper perty and to the nether . whych schewyth hym opynly in the middyl in þe bowȝt of the arme þat ys for he growyth of the schuldre veyne a boue . and of the arme hole . by neth as yt semyth opynly . ¶ Therfor of al the bonys of on . of whych ys hadde knowlege . in nowmbre ben . ȝi . and þat ys . yf we make departyng by twen*e* þe rall f. 24r and the adiutorye . wet thow wele þat all þe vaynes þat cu*m*myn from þe lyu*er* a bouyn*e* ben*e* dep*er*tyd fro the lesse branche of the gret vayn þat co*m*myth þat cu*m*myth fro the lyuer . and þat vayne ys deuydyd into iij branches . of whych oone entreth þat goyth to the mydreste . and other gooth to þe chest . of the hert . and yn hys oute goyng from þat place . he ys dep*er*tyd . *and* oo p*er*tye of hym ys knit to the ryght eere of the hert . and þis branched ys dyuyded in to the branches whych oone entreth the holownesse of the herte . ¶ The ijde ys spred vppon the vpper p*er*tye of the herte . ¶ The iijde ys sprede vppon the nether p*er*tyes of the brest and to the forcle of the brest . And fro f. 24v þe branche of the forcle grouyth a branche whych goth to þe schuldre and to the arme hole . and other branches goyth by the inn*er*mer p*er*tyes of the throte to the braunche of þe hede . and to þe vpper membres . and þ*er* be fastned that nedyth . And yt schewyth whych ys the kyndlyhed of þe fosayd venys w*ith* þe hede þe herte and þe lyuer after thys man*er* dep*er*tyng .

Focilles . racet . Cephalica . Finis brachij . Basilica . Salvatilla . or Epatica . or splenatica . Humerabilis . Sepha-

[13] **710** the] *del.* lyu*er*

lica . or Assellata . Purpurea . or Nigra . or Communis . or matrix[14]

f. 25r The iijde chapithle of Anothomie and shape of þe forchle *and* þe rybbys *and* of þe chyne bone . *and* of þe brest on þ*at* p*er*tye vnto þe mouthe of þe stomake . Under the throte yn the forparty ben*e* ordeyned the . ij . bones of the forcle . whych ben*e* rounde wyth oute . and holowe wyth yn . In the wych the ende of þat oon*e* ys fastened yn the sholdre . And the oder ende ys fastened wyth vppermor partye of the breste . where ys the pytt of the throote . Vnder whych forcley and boxe or pyt ben*e* ordeyned the vij bonys of the brest yn the ende . Of the whych . vij . bones . ys a grystyle . and yt lastyth the length of these bones wyth þe fastnyng of hem frome the pyt of þe þrote . f. 25v vnto the lyuer vnder the tetys . and þ*er* yn that place or nygh be neth ys þe mouth of the stomake or the holownes of the brest And þis length wyth the knettyng to gydd*er*s ys clepyd . thorax . The makyng of þese bones wyth rybbys and rygges . ys clepyd p*roper*ly the brest . ¶ Ther ben þen xij . rybbys whych ben fastened w*ith* xij spondelys . and þese spondelys ben clepyd p*roper*ly the spondys of the brest . the remnaunt of the spondels that ben . v . ben clepyd the spondels of the reynes . whych . xij . rybbys ben bowed yn the maner of a halfe compas . of the whych . vij . of the hynd*er* endys ben fastned wyth the spondelys of the bake . And the grettest of þem ys yn f. 26r the myddys . but the former ende of them ben fastened wyth the vij bonys of thorax . and v . of þoo xij rybbys ben*e* mayd schorte . and be neu*er*more fastened wyth thorax . but w*ith* the spondyls . behynforth . And so þei be clepyd hynd*er* rybbys . For why þat whan the

[14] **751** matrix] *bot. marg.* the thyrde

former endys ben*e* bowed downe . þey bowen vpward aȝen*e* forthy that þei haue no fastnyng . os haue the vij grete that be fastened w*ith* the bonys of the brest . and takyng here knyttyng of hem wyth þese bonys[15] of the brest and the holownese of the rybbys . And yn the spondyls of the brest ys sett the hert . and yt boueth aft*er* hys beyng moyst to the lefte syde . and the smale ende of hym boueth more nyȝ the ryght syde . ¶ And the lung ys sett yn the same holounesse f. 26v . but aft*er* hys beyng he loweth to the ryght syde . and yn the myddys of hym he boueth much to the left syde . And þer ys also a welle or a skyn of synewes . whych ys of the kynde of the mydryfe wonden to gydd*er* of synews of gret veynes . and smale as the same mydryfe . and dep*er*tyth the brest in two . The lung and the brest he ys fastned w*ith* the xij hynd*er* spondels of the brest . And thys dep*er*tyng of the brest was doo for nothyng els but þat ȝyf that one halfe of the lung drauyng as kynde askyth . wyth þat other halfe in drawyng of eyre and outcastyng myght helpe hym selfe . and þis makyng was kyndely for the noble hed of þe worchyng . f. 27r And the wertu of thys and hys nede fro þe fyrst spondyl of the brest . whych fyrste spondyl of the brest ys the ayth of all the spondyls . begynnyng at the begynnyng of . Nucha . fro þat place a forseyd comen*e* two synews of mouyng and felyng of þe brest and of . Thorax . Off whych ben*e* mayd the muscles . and þe mouyng synewes of the forsayd brest . ¶ And þou schalt vete þat þe muscles and the mouy*n*g synews of the same brest . sum ben*e* wylfully mouyng . and þei com from the more partye from the vjth payre of synew-

[15] **794** bonys] *dup.* bones

es of the brayn of Nucha . and su*m* synew-
es mouen by the kynd of the makyng
f. 27v of the brest . and þat ys wyst by the condic*i*on
of the apoplyce . by whych the brest moueth
to sore the tyme of sekenes . and yt moueth
not yn the oure of sekenes by the vertue co*m*-
myng fro the brayn . whan al the brayne
ys funde stoppyd . and þat oure soth vertues
wyth the spyrytys of the soule may not goo
down by the senewes .

The veynys comen to þese kynde de-
p*er*tyng . the brest . ben*e* branched of þe
ijde branches of the veyne of gret vertue . hauy*n*g
hys begynnyng yn the rowndenesse of the
lyuer . whych branche co*m*meth to the myddrefe .
And from the myddrefe he goyth to þe skyn .
dep*er*tyng the brest yn the myddys endelong .
¶ And wyth þese co*m*men other veynys fro*m*
f. 28r the branche þat ys the iijde . of the vnknown
veynys yn the ryght ere of the forsayd hert .
¶ Sothly the art*er*yes þat co*m*men to the
forsayd skyn and to the brest . co*m*men by b*ra*n-
chyng of a branche . dep*er*tyng from a grete
arterye . the whych growyth in the lyfte ere .
of the herte . And eu*er*y art*er*ye passyth from þe
branchyng to eu*er*y veyne . þat the membres
whych ben*e* noryshyd of ony veyne of veynes
ben*e* quykkyd and kepte of the arteryes pas-
syng vnder the same veynes .

Thow schalt wyt þat all the arte
ryes and the veynes whych co*m*men
openly to norysch and to quykkyn þe mem-
bres of the brest and of torax . and specyally
the grete senewes co*m*myng to the makyng
f. 28v of the muscles of the brest . hauyng her begynny*n*g
of Nucha . or of the spondels of the brest . wher-
fore yt schewyth opynly how and yn what ma-
ner þat þe apostumes of þese places schulde be
kut . ¶ And how art*er*ies owen to be don*e* and þe
places yf yt be nede . for yt acordyth þat hys wyr-
chyng schal be don*e* . Aft*er* the long rybbys for þe

more party . And schal þ*er* be no dysseyte ne no hurtyng of the synewes . ¶ After the[16] contrarye of hys goyng whych ys ful goode and p*ro*fitable for the lowest and vttermost p*er*tyes . Suche cuttyng ȝeldyng or makyng moyste fayrest dryeng . and moyst acordyng and spedy aft*er* the goyng forth to the membres os yt schewyth openly .

The iiijth Chaptile of the Anothomie of the vtter wombe from the mouth of the stomake vnto the hanches and the spondels of that p*er*tye . f. 29r

Affter the rybbys yn the forpartye *and* the vttermore partye ys a skyn keu*er*yng the stomake and the guttys . and ys fastened w*ith* softnes vnto the luine of the share . All thys place ys neshe vnto þat p*er*tye . and thys vtt*er* skyn ys clepyd the wombe . And vnd*er* thys skyn ben ordenyd muscles acordy*ng* to thys place . and þei ben viij . in noumbre . ¶ And now I schal tell to the makyng of hem after þese musclys . yn the for p*er*tye . ys ordeyned a greyned skyn hauyng wyth w*ith* hym þe cal of the guttys . of the brekyng of f. 29v whych greyned skyn whych ys clepyd Cyphac . ther fallyth the goyng oute of the ȝerbus . of the gutt . to the ballok codde . and oþ*er*whyle þ*er* ys mayd swellyng yn the flank . only for the goyng . And þan wetyn lechys þat the brekyng of the syphac . ys not gret . whych schal lyghtly be helyd wyth playstres and restyng wyth lyggyng vp ryght and vnder lyggyng the body . ¶ Vnder þis Cyphac ys put the bone of the schare yn man*er* of an halfe compas saue þat yn the ouer p*er*tye toward the nauyl he hath som goyng . *and* ben*e* yn the shares wyth the hanches . and þ*er* he ys sundry goyng . and ben tha he may susteyn tho partyes wyth hys hardnesse . wyth þe

[16] **872** the] *del.* goyng

f. 30r byndyng of the whych growyth oute a synewe bodylyk to the byndyng of a rope . Vnder whych ben*e* many arteryes and venys after thys gretnes askyth . ¶ And thys membre synew or arteryousse so Imayd . ys clepyd a mannys yarde . In the hede of whych ys fleshly fulfylly*ng* . and yt ys coueryd w*ith* a skyn . þat he be kept fro*m* hurtyng wyth oute . And þat the rubbyng of the skyn of the hede there be gettyn more apetyde yn delyng or lyggyng w*ith* women . and that of whych rubbyng the sperm may bett*er* be cast oute[17] . yn to the vessyl of yt by swellyng of the ȝarde . castyng or mouyng of[18] the skyn vpon the hede of the ȝerde . he castyth out bett*er* the sperm . ¶ Ferthermore from the Cyphac gone down ij pypys . of the whych the f. 30v lower pypes ben mayd wyde . and þ*er* the cotys of the ballok and the skyn growen or bene brede . yn whych the stonys be contenede . ther comen whyte that from the p*er*tye of þe femorayllys to the p*er*tyes of the ballokys branches and veynes . whych gon*e* down*e* from the veynys many folde in lapped . and þei taken substance of whytte and glandulouse flech . whych flech what eu*er* the blode be yn the veynys of yt . yt ys chaungede a none . and mayd whyte . and þan þat flech . puttyth yt forth to the ballok . and þer ys mayd p*er*fyte sperme . ¶ Aft*er* ther be gendred ij holys whych goon to the yerde . of whome the ryghttyng or standyng ys made . whan the vessels and þe f. 31r gret holys of yt ben*e* fulfyllyd . w*ith* the whych spyrites and hete . and þe vessyl of sperme be mouyd to the multytude of sperme . or els to the sharpnes of þe same sperme . whych ammouyng sesyth not tyl the sperme be cast oute . ¶ Sothly the vpper reme or skyn*e* whan yt ys strechyd aft*er* the place

[17] **921** oute] *del.* And that the rubbyng
[18] **922** of] *dup.* of

of the flank . and dindimus *id est* the vnder
skyn a boute the balok ou*er* the pypys þat
co*m*meth from the Cyphac . ys mayd the
vttermore pouris of the balloks .
In the woman for sothe in þe stede
of the ȝarde . ys the neke of the mo-
der . and þat matrix ys made synowy . þat
yt may be mayd strayte oute . And mayd
large yn tyme of beryng . whan yt ys ne
deful . And yt hath wyth þat ij echyng . and f. 31v
yt hath smalle balokkys neþ*er*les brode . and
the neke of yt in comp*er*y*ci*on to þe ballokkys
ys yn man*er* of a ȝerd or a pyntyl . y turned
inwarde . and the hede of the neke yn tyme
of castyng of sperme tocheth the ballokys .
and mouyth hem . þat þei castyn þe better
the sperme to þe holownesse of the maryce .
and from the hede of the neke yt ys bore forth
yn to the schape of a woman . and from then*es*
yn that place yt entreth ynto the schape yn
to the neke of marice . the ȝerde of a man yn
tyme of goyng to gyddyr . the matrix ys p*roper*-
lyche set betwene the gutt clepyd langaon .
and bleddyr . and ys bound wyth the lud-
dokkys . and wyth the chynne bones w*ith* loosse
byndyng and not strayte . þat yt may be mo- f. 32r
uyd a*n*d made more large yn tyme of conce-
uyng
Affter þe ende of þe xij . rybbys ande
þeier lyṇkys or her spondels yn the
neþ*er* p*er*tye ben ordeyned . v . linkys or spondels
of the reynes . and þei ben fastened yn the
neþ*er* p*er*tye towarde the ars . wyth the last bone
the whych bone semyth to be mayd of iij p*re*-
uy bones . þat ben lyke to the schap of the
lynkys or spondels to the whych last bone
þe grestely bone of the ars ys knyt . the whych
ys mayd of iij p*er*tyes yn hys last ȝarde . par-
tye ys grystely sympely . from eu*er*y hole of þese
spondels ys brede or brought forth a synew
from the bone of the tayle or of the ende of þe

f. 32v chynne goth oute oon*e* synew . for yt hath but oon*e* hole . and the synews co*m*myng to the makyng of the muscles of the vtter wombe and þe branchyng veynes of the grett*er* go-yng downe by the ij branches of þe reynes . and that o*þer* of the more veyne . dep*er*tyth from the lyu*er* or from the branches of hem . by the bak and the vtter p*er*tye to the noryshyng of þe membres and musclys brawnys of þe wom-be . ¶ And after þe more arterye goyng down to þe myddryfe co*m*men braunches to þese . vntt*er* partyes and muscles of the wombe . whych quykyn*e* þoo places wythinne thys ho-lownes ben*e* set yn the noryshyng membres . and vppon þese noryshyng membrys and vn-der þese spiritual membris . ys a new mem-[19] f. 33r bre wouyn and knyt of muscles and artery-es . and yt ys mouyd of mouyng . and drau-yng oute of vynde . and yt ys departyng of the noryshyng membrys fro þe spyritual . and yt ys clepyd diafragma . þ*at* ys i*n* englych the myddref . and he holdyth hys place lyke the schape of a bely yn that worchyng . ¶ And whan þe membre ys[20] hurt what for þe makyng of hym . and what for hys nedeful and p*ro*fitable werchyng continuel by þe wonde syde to the cural and dedelych as the most partye vnd*er* the myddryf yn the ryght halfe a sell naturally the lyuer ys knyt and bounden to a lytyl nett . whych ys rote of al veynys w*ith* ynforth and wyth-outeforth . and he hath hys holownese f. 33v yn the myddys of hem . ys þe chyst of þe gall or þe bagg of the colrem . of whych ij vessyls or ij . pypys ben dep*er*tyd of whych on*e* beryth colrem to comfert þe dygestyon yn þe sto-make . and þat o*þer* beryth coloure to the gutt þat þe castyng oute be comforted *and* þe slymehed be remouyd fro þe ordure .

[19] **1006** mem-] *bot. marg.* bre wouyn
[20] **1014** ys] *del.* hynt

In the left halfe forsoth ys set the melt . and yt ys bunden to and fastned w*ith* þe lyu*er* goyng be twyxte veynes and weyes by whych þe malycolyus sup*er*fluytes ben ledde to þe splene . *id est* . the mylt . and þat splene sendyth oute by dew weyes . sum of that sup*er*fluytes to the mouth of þe stomake þat yt co*m*forth þe apetyte . and sum partys of hyt sendyth to comforth þe dygestyon w*ith* holdyng yn [f. 34r] the gutt yn the stomake . ¶ In the mydward of þese membres ys the stomake . whose vttermore partye ys strayt and narrow . In hys neþ*er* partye ful large . And þe[21] neþ*er*mer partye ys endyd yn þe place of the nauyl . and ys clepyd the fyrst vesyl . in whych nature makyth fulfyllyth fyrst hys dygestyon kyndely w*ith* þe bottom of the stomake . yn the wych p*re*ncypallych þe dygestyon vaxeth and thryueth and ys noryshyd . ¶ Ther to the stomak ben*e* tyed and fastned . and contenyd dyu*er*se gutt . whych aft*er* þe deference of her places and steedys are dyfferenced of whych gutt yn nowmbre ben vj ¶ The fyrst ys clepyd duodenu*m* . for þi . þ*a*t [f. 34v] ys yn length aft*er* the mesure of xij ynches . ¶ The ij and þe nexte ys clepyd ieiuniu*m* or the holow gut . or the fastyng gut . and yt ys clepyd so . for yt ys eu*ere* founden voyde . and empty . And yn thys specyally and yn duodenu*m* . yn the botum of the stomake vnset and fycchyd clene and sotel veynes . by the whych þe lyuer draweth to hym the clennesse and the best of mete and þe drynke defyed . and membres casten oute the same clennesse to þe lyu*er* . ¶ wyth ieiuniu*m* ys knyt yllyon . that ben*e* smale gutt . After hym ys ordenyd monoculu*m* . and yt ys clepyd so . for yt hath none hole butt

[21] **1044** þe] *dup.* the

one yn þat one ende . wyth þis ys fast
f. 35r ned colon . or the rownde gutt . wyth colon
ys fastned longaon . and he goyth ryght
on sum spondels of þe reynes . and ys ended
yn the ars . and þer ys[22] euyn ronny*ng*
the hole of þat ars . gret veynes to whych
the lyu*er* and the mylte cast oute moche ma-
lyncolyose blode . and whan þese veynes are
opyn*e* . þei ben*e* clepyd the pyles and þe eme-
roydes . ¶ Vpon*e* þis gut yn aman toward
the share . and yn a woman towarde þe ma-
ryce . ¶ Toward the share ys set þe bledd*er*[23] .
whych ys þe vassyl of the vryn . and yt ys
synewy . and the neke of yt ys fleshy os
oft nost . and she hath yn hyr two cot*s* the
whych ben*e* ij . skynnes . and þ*er* be yn her ma-
ny smale veynys and art*er*ies . of whych
f. 35v yt purchesyth and noryshyth þe lyffe . And þe
neke of yt passeth forth vnd*er* the bone of the
share . and yt ys mayd smale in his goyng
forth to þe vtt*er*more partyes . And the wey
of yt ys mayd yn the man*er* of a grete veyne .
yn a man*e* . and entryth þe makyng of the ȝar-
de . and passyth forth yn to þe fleche byneth
forth . And partyth yn man*er* þe flech of the
ȝarde by the myddys . and by þat party goth
vnto the vryn . ¶ Ther ben*e* yn the ȝarde
of a man ij holys at the left . one ys by the
whych þe vryne ys cast oute . and þis ys
fastned w*ith* the neke of the bladder . And a
noþ*er* by whych kynde of a man ys caste
oute . and þis ys fastned wyth the kynde
of the vessyl of sperma . In þe holownesse
f. 36r of the ȝarde þese ij holes ben*e* mayd as oon*e* .
Soth yt ys as sum seyn þ*a*t þ*er* ys þe thyrde
hole by the whych ys cast out sperme or
sede yn slepe . and yt ys dyfferensyng fro*m*
the tod*er* . but yt ys vnknown to me .
In the bladder of a[24] woma*n*

[22] **1074** ys] *del.* ended
[23] **1081** bledd*er*] *out. marg.* the bladder

ys nought sought ne purchesyd for yt ys ful schort yn þis man*er* and strate . and the kynd of her ys not cast oute yn thys man*er* . ¶ And for the schortnesse of the nature ys myghtty kynde ys cast out yn þis man*er* lyghtly of al sup*er*fluyte wyth holdyng yn yt . of whych myght be beryed þe stone gret and smale yn the ende of the xij . spondels of the rybbys vnd*er* the fyrst spondel or lynk of þe reynes ben*e* set ij reynes þat ys f. 36v ij kydnerys yn the nere party of þe body . and þe ryght ys hyer þan the left of holownese of whych ben*e* departyd ij pypys . and þei go-down by nethe . for tyl þei be ioyned wyth þe vtt*er* skyn of the bladder . And þei entren aft*er* þat þe nek of the bladder . and go down*e* and p*er*chyng þat other skyn of the bladder . ¶ And by þese veynys and aft*er* þis maner þe vryn passyth from þe reynes to þe bladder and þese veynes ben clepyd . vriclides . and yt ys opyn knowyng þat þe passyng forth of the synews co*m*myng from the spondels of the reynes and veynes and arteryes gone forth almost into the utt*er*more wombe . *and* the flank aft*er* the granes of þese places and therfor yt ys gode for to knawe þe kuttyng f. 37r of the places of apostumes aft*er* the greynes . And the cauteryes owen to be don*e* aft*er* þat maner of her places .

The vth capitle of Anothomie and þe shape of the haunches of the þyes *and* of þe ha*m*mes . of the sky*ne* and of þe fote . of the toes . and of the membres beyng wyth yn the*m* .

The bones of þe haunches ben*e* ij . and þat ys to say . one yn the ryght halfe . a no*þer* in the left halfe . and eche of hem ys fastned yn hys partye wyth the bone of the ryge . the whych bone ys mayd of thre partyes the

[24] **1108** a] *del.* man

whych bene lyke to the spondels of þe reynes . ¶ In that p*er*tye the bone of þe haunche f. 37v hath a grystyl w*ith* þe man*er* of a brodnesse . In þat o*þer* ende hyt hath an opyn gretnesse . yn the whych ys an holownesse clepyd the box of the hanch yn the whych ys fastned the rounde ende of the bone . And þis place ys bonden stronge insencyble bondys or vnable to be felt . the whych ben*e* cordy synews . And yn þe myddys thys box ys bownden þe whych helpyng or goyng the whylrbone w*ith* the box . whych bo- ne whan*e* yt ys brosten or dystroyd þe whych bone goth oute of hys place although yt be restoryd of ache by a leche þe ioynyng nou- ght . but steryth oute . and yt ys no helyd so þat the pacyent schal halte al hys lyffe tyme . ¶ In þe token of the softnesse of þe bone ys þat . whan the whyrlbone ys res f. 38r toryd þe p*er*fyte restoryng dureth not . ne ys not fastned . but yt hyppyth out . ¶ After þe box ys ordenyd þe bone of the thyȝ . þe whych ys a marowy bone rounde . and entryth the box w*ith* oute and w*ith* yn holowe . whose vppermore ende entreth the box of the haunche . and yt ys ful rounde . And yt ys bounde as we seyde a fore . and þ*er* ys turned a boute yn þe tyme of mouy*n*g of the legg and of the thyz and of þe foote . ¶ Forsoth þe nether ende entryth the box of þe more focyle of the skyn . and yt ys fast- ned wyth þe ij skynne bonys . and yt ys bon- den wyth strong bondys vnsencyble . And þis ys done yn eny knotty ney*þer* mouyng ne rubbyng þat membre shal fele no dise- ase ne hurt . ¶ Vpon*e* þis Ioynture the f. 38v whych yn the kne ys put arounde grystle bone þat ys clepyd the Iye of þe kne . and of sum þe appyl of the kne . Vnd*er* the kne ys ordeyned the bones of the legg . the whych ben*e* marown . And þei ben*e* contenyd and fastned aft*er* the length to þe ancle and the heyle . as yn plantys . and þei ben clepyd ij focylles of þe legg . the more and the lenger and þe gret*ter* ys clepyd þe nether focill . And þe lasse and þe smaller bone . and yt ys þat þe more schort*er* bone . and yt ys clepyd þe vpper focyller . and yt makyth the sperler of the legg . ¶ And þe ende the ij bones hauen in þe nether ende ij echy*ng* and þei ben*e* fastned to þe bone of the heyle . to the whych ys ioned a boue ys clepyd nauicula . of the neþ*er* p*er*ty ys aboue assub f. 39r to whych iij bones þat ys to þe bone of the heele nauicula . and assub ys ioned to the rasche of the fote . whych aft*er* som ys made of iij bones . and aft*er* sum iiij ys þe trewer . Sothly þe rachett of the fote ben*e* set to and fastned to þe instape . the whych ys mayd of v . bones of þe stape of the fote be fastned þe bones of þe tooes of the feete . þe whych bones of the tooes of the fete ben*e* nowmbred . xxiiij . for yn eu*er*y too of þe fote . ben iij bones safe only þe gret too . whych ben*e* but ij . bones forthy þat yt nedyth no gret mouyng . ne yt ys not necessarye for þe schap of þe toes whan þei folde . for the toes aske not such foldyng . al yt were nedeful be that yn the thombe of the f. 39v honde þat yt haue iij . þ*er*for as yt shewyth the bonys of al the fote wyth the hanche from þe one p*er*ty ben*e* . ȝi . From þe holys of þe v . spondels and þe holys of þe last bone ben*e* dep*er*tyd synews . the whych ben*e* branched and sum be þe hynd*er* partye to þe hanche aft*er* the length . and þeuen to þe places felyng and mouyng and wylful steryng . and þese ben*e* medlyde yn that place w*ith* branches . and þe flech . and ar mayde aft*er* þat place yn þe legg . from whych þei ben dep*er*tyd from eu*er*y ende of the corde . mouyng þe neþ*er* parties and þe vpper parties and specyally þe kne and the skyn . and þe legg . drawyng hem Inne and oute . And all

f. 40r the synews bondys gon*e* forth aft*er* þe lenght of þe thyȝ and shynne . Ther ben*e* made also braunches of þese synews co*mm*yng from the holys of the spondels of þe reynes . and fro the hole of the last bone grete muscles and smale yn the legg from the endys of whych mouyn þe neþ*er* parties . and þe neþ*er* membris as þe foete . and the toes . And þ*er* be mayd yn the kne yn the apple and yn þe sydes of yt . and yn the calfe of þe legg by kynde . of whych su*m* maky*ng* of noble synews and braunches þorough hepbe of whych syneus or muscles for the knyttyng to gydder of hem whan þei be wounded or prykkyd . þei bere a kyng . and dyse f. 40v ase to þe rotys of þem so and yn suche man*er* . þat the leche may by no craft . ne by no suttylte bryng to remedy . but þat the seke scholde dye . And forthi þe wondys of þis place be ful dredeful . and vncurable for þe most p*er*tye . for the ende . of þe vttermore partyes of þe veynes . and art*er*ies gown*e* down*e* fro þe reynes gown brau*n*ches . whych braunches or veynes ben*e* departed fro þe more braunches of þe veyne . hauyng hys bygynnyng from þe lyuer . and the braunches of þe arteryes ben*e* departed of art*er*ies comyng fro þe more braunche of the ryght ere of þe herte . and of þe myddryfe gon*e* downe veynys by þe bak and be medled wyth veynys[25] f. 41r knyttyng and comen to gryndes and to the haunches . and gone down wyth þe musclys and syneuys aft*er* the length to þe ende of þe toes . And of al these thyngs ys mayd the vttermore sencyble skyn .

¶ Of þese membrys ys noryshyd and quykkenyd . and þ*er* fore yt shewyth opynly how the knyttyng of þe apostumes of

[25] **1262** veynys] *bot. marg.* knyttyng

þese places owen to be kut aft*er* the length of the legg and of the thyes . and vppon the henches aft*er* þe length of þe spondels and the cuttyng own*e* to be down after the length of þe body . and þe caut*er*ies of þese placys on þe same man*er* after the goyng forth . In þese membrys . Oroniam membrox hu- f. 41v mani corporis ponens om*m*ino ignorate erant . placuit ueteris medicis maxime galieno . ut p*er* anathomias brutoru*m* . Alii int*er*ior membros po*n*nes manifestarentur . ¶ Cum aute*m* su*n*t bruta animalia nobis queda*m* su*n*t similia . ut semie . Sod*er*in uero membros cam exterioru*m* qua*m* int*er*ior nulla nobis adeo i*n*ueniuntur similia ut porci . ¶ Ideo i*n* eis anathomia fit . Est autem anathomia equalis membros que sic fit . Porcu*m* in uersum debes ponere :

CHAPTER FOUR

THE GLOSSARY

4.1. Conventions

The present section provides a complete glossary for the words appearing in W290 (ff.1r-41v), offered as a tool kit for an efficient reading of the text. Each entry has been conceived to display the following information: the headword, the part of speech, the meaning of the item, the different allographs for each item with their corresponding number of occurrences.

Given the inconsistency of Middle English, headwords have been highlighted in bold type under a uniform spelling, based on the most frequent spelling of each word in the witness. These headwords have also been lemmatised using the online version of the *Middle English Dictionary* (henceforth *e-MED*) (Lewis et al. 1956-2001).

Next, the corresponding part of speech appears italicized immediately after the headword, employing the following notation: nouns (*n.*), pronouns (*r.*), deverbal nouns (*ng.*), adjectives (*a.*), adverbs (*b.*), verbs (*v.*), anomalous verbs (*va.*), determiners (*d.*), conjunctions (*c.*) and prepositions (*p.*).

The meaning of each entry is partially taken from the *e-MED*. If a word is not included in the *e-MED*, other sources, such as the *OED*, have been consulted. Finally, the different allographs appearing throughout the text are listed under each entry, together with the number of occurrences for each of them.

The Latin terms appearing in the text have been disregarded in this glossary. Capelli's *Dizionario di Abbreviature Latine ed Italiane* (1990 [1912]) has been very helpful in the transcription of these items.

4.2. The glossary

A

able, *a*. Of things: suitable for a use or purpose
able (1×)

abyden, *v*. To remain firm, solid
abyde (1×)

accorden, *v*. To be fitting or proper
accordyth (2×),
acordyth (1×)
acordyng (2×),
acordy*ng* (1×),
accordyng (1×)

ache, *n*. Pain
ache (1×)

addicyon, *n*. Addition
addicyons (1×)

adiutorye, *n*. The bone of the upper arm (humerus) or the bone of the thigh (femur)
adiutorye (13×),
adyutorye (1×),
adiutorie (1×),
adiutorye (1×)

akyng, *n*. Pain, aching
akyng (2×)

alauda, *n*. A bone of the head
Alauda (4×)

aliuelych, *a*. Alive
aliuelych (1×)

almost, *b*. Very nearly, almost;
almost (1×)

alteracion, *n*. Change, transformation
alterac*io*n (1×)

alteraten, *v*. To change, modify
alt*er*atyd (1×)

althermost *b*. Most of all
althermost (1×)

although, *c*. Although
although (2×),
althowgh (1×),
Allþougth (1×)

anathomia, *n*. The science of anatomy
anathomia (2×),
Anothomie (2×),
Anothomie (1×),
anotome (1×),
anathomias (1×)

ancle, *n*. Ankle
ancle (1×)

anenste, *p*. Next to, close to
anenste (2×)

anoþer, *d*. Another
anoþ*er* (6×),
another (4×),
anoder (1×)

aperten, *v*. To open
apertyth (1×)

apetyte, *n*. Appetite
apetyte (1×),
apetyde (1×)

apoplyce, *n*. Apoplexy
apoplyce (1×)

apostume, *n*. Any morbid swelling or inflammation in any part of the body, external or internal
apostumes (3×),
apostomes (2×),
apostummes (1×),
apostomys (1×)

apple, *n*. A dry-scabbed abscess on the leg
apple (1×),
appyl (1×)

applien, *v*. To join or combine
applyed (1×),
aplyed (1×)
arme, *n*. Arm
arme (17×),
armes (1×)
ars, *n*. The anus
ars (4×)
arterie, *n*. A blood vessel of the arterial system
arteryes (17×),
arterye (4×),
art*er*ies (4×),
art*er*yes (2×),
art*er*ye (2×),
at*er*ies (1×),
artyres (1×),
arteries (1×),
arteria (1×),
arterys (1×)
arteryousse, *a*. Having characteristics of an arterie
arteryousse (1×)
asken, *v*. To ask
askyth (2×),
aske (1×),
askyng (1×)
aȝene, *b*. Again
aȝen*e* (1×),
aȝen (1×)
ayth, *n*. A harrow
ayth (1×)

B

bagg, *n*. A bag
bagg (1×)
bak, *n*. Anatomy: back
bak (2×),
bake (2×)
baker, *n*. A baker
bakers (1×)
ballok, *n*. A testicle
ballok (3×),
ballokys (3×),
balokkys (1×),
balok (1×),
balloks (1×)
begynnen, *n*. To begin
begynnyng (5×),
begynny*ng* (2×)
behaven, *v*. To conduct or comport
behaue (1×)
behynforth, *b*. Along or at the back
behynforth (1×)
bely, *n*. A belly
bely (1×)
ben, *va*. To be
ys (323×),
be (102×),
ben*e* (73×),
ben (41×),
is (10×),
was (7×),
were (3×),
ware (2×),
are (2×),
bene (2×),
beth (1×),
ar (1×),
beyth (1×),
beyng (8×),
beyng (2×)
beren, *v*. To carry (sth.), bring
beryth (3×),
berer (1×),
bere (1×)
beryng (4×)
berien, *v*. To bury
beryed (1×)
best, *a*. Best

best (2×)
better, *a.* Better
better (7×),
bett*er* (3×),
bettyr (1×)
betwixte, *p.* Between
betwyxte (2×)
blad, *n.* Something shaped like a blade
bladys (1×)
bladder, *n.* A containing organ or bag (the urinary bladder)
bladder (6×),
bleddyr (1×),
bledd*er* (1×)
blake, *n.* The pupil of the eye
blake (2×)
bledyng, *n.* The action of bleeding
bledyng (1×)
blode, *n.* Blood
blode (7×)
body, *n.* The human body
body (4×)
bodylyk, *a.* Physical, corporeal
bodylyk (1×)
bond, *n.* Something used for tying
bondys (3×),
bondes (1×)
bone, *n.* A bone
bone (69×),
bones (34×),
bonys (23×),
boyn*e* (1×),
bon*e* (1×),
bon (1×),
bone (1×)
bordon, *n.* A shield
bordons (1×)
bore > **beren**, *v*
bottom, *n.* base, foundation, bottom
bottom (1×),
botum (1×)
bouen, *v.* To change direction or position; To bend, curve
boueth (3×),
bowed (3×),
boyth (1×),
bowen (1×),
bow (1×)
bounden, *v.* To tie
bonden (4×),
bounden (2×),
bunden (1×),
bynden (1×),
byndyth (1×),
bownden (1×),
bounden*e* (1×),
bounde (1×),
bondyn (1×),
bound (1×)
bowȝt, *n.* A limb
bowȝt (2×),
bowht (1×),
bouȝt*e* (1×)
box, *n.* A socket (of a bone)
box (11×),
boxe (1×)
branche, *n.* A branch-like extension or part, a ramification
branches (8×),
braunches (6×),
branche (6×),
braunche (2×),
brau*n*ches (1×)
branchyng, *n.* Of blood vessels, nerves: ramified
branchyng (2×),
br*a*nchyng (1×)

branchen, *v.* Of blood vessels, nerves: ramified
branched (3×),
branchyd (1×)
brawny, *a.* Made of muscle
brawnys (1×)
brayn, *n.* The brain
brayn (31×),
brayne (6×),
brayn*e* (2×),
brane (1×)
breden, *v.* To spread
brede (1×)
brekyng, *n.* The act of breaking
brekyng (5×)
brest, *n.* The front of the thorax, breast
brest (31×),
breste (1×),
brest (1×)
brethen, *v.* To breathe
brethe (1×)
brode, *a.* Wide, broad
brode (8×),
broode (1×),
brede (1×)
brodnesse, *n.* Width, breadth
brodnesse (1×)
brosten, *v.* To be broken
brosten (1×)
brought > **bryngen**, *v.*
bryngen, *v.* To bring
bryng (2×),
bryngyth (1×)
but, *c.* But
but (23×),
butt (1×)
by, *p.* By
by (59×)
bygg, *n.* A teat
bygg (1×)
bygger, *a.* Bigger
bygg*er* (1×)
byll, *n.* A cutting, hacking, or grubbing implement
byll (1×)
byndyng, *n.* The act of tying
byndyng (15×),
byndy*n*g (2×),
byngdyng (1×),
bynedyng (1×)
byrd, *n.* A fringe or stripe
byrdys (1×)

C

cal, *n.* An enveloping membrane
cal (1×)
calfe, *n.* The young of cattle, a calf
calfe (1×)
casten, *v.* To throw
cast (7×),
castyth (1×),
castyn (1×),
casten (1×),
caste (1×)
castyng, *n.* The act of throwing
castyng (3×)
cause, *n.* Cause
cause (7×)
cauterye, *n.* A cauterizing operation
cauteryes (1×),
caut*er*ies (1×)
cauteryenge, *n.* Cauterization
cauteryenge (1×)
cercle, *n.* A circle
cercle (2×)
chanel, *n.* A canal
chanel (12×),
chawel (1×),
chanells (1×)

chaungen, *v*. To change
chaungede (1×)
cheke, *n*. The lower jawbone
cheke (3×)
chest, *n*. The chest
chest (1×)
chewen, *v*. To chew
chewyth (1×)
chynne, *n*. The chin
chynne (2×),
chyn (1×),
chyne (1×)
chyst, *n*. A sac or bladder
chyst (1×)
clennesse, *n*. Cleanliness
clennesse (2×)
clensen, *v*. To cleanse something or filth or dirt
clenyth (1×),
clene (1×)
clensyng, *ng*. The act of cleanse something
clensyng (1×)
clepen, *v*. To apply a name
clepyd (59×),
clepid (2×),
clepyde (1×),
cleped (1×)
closen, *v*. To close
closeth (1×)
clothy, *a*. Covered or made with clothes
clothy (2×),
clothelyk (1×),
clothelych (1×)
cloþen, *v*. To put clothing on
cloþed (1×)
cloyth, *n*. Clothing
cloyth (3×),
clothes (1×)
clysterye, *n*. A clyster, enema
clysterye (1×)
codde, *n*. The larynx, throat
codde (1×)
colon, *n*. The large intestine or colon
colon (2×)
coloure, *n*. Colour
coloure (1×)
colrem, *n*. Bile
colrem (2×)
comforthen, *v*. To strengthen the functioning of an organ
comforth (1×),
comforted (1×),
comfert (1×),
co*m*forth (1×)
commen, *v*. To come
co*m*men (7×),
co*m*meth (4×),
com (4×),
comen (3×),
co*m*myth (2×),
cu*m*yth (1×),
cu*m*myth (1×),
cu*m*en (1×),
comen*e* (1×),
cu*m*myn (1×)
commynge, *n*. The act of coming
co*m*myng (14×),
cu*m*yng (1×),
comyng (1×)
comone, *a*. Common
comon*e* (2×),
co*m*mone (2×),
co*m*e (1×)
compas, *n*. A circumference
compas (2×)
comperycion, *n*. Comparison
comp*er*yc*i*on (1×)
comprehenden, *v*. To contain or

include
co*mpre*hendyth (3×),
comp*re*hendyth (2×)
comprehendyng, *n.* The act of containing or including
comp*re*hendyng (1×)
conceuyng, *n.* The act of conceiving
conceuyng (1×)
condicion, *n.* A situation or state
condic*i*on (1×)
consyderen, *v.* To inspect, examine or observe
consyderyd (1×)
contenen, *v.* To contain
contenyd (2×),
contunyed (1×),
conteyned (1×),
contenede (1×)
continuel, *b.* Continually
continuel (1×),
co*n*tynuel (1×)
contrayie, *n.* Something located on the opposite side of or over against (sth.)
contrarye (1×),
contrary (1×)
contynuen, *v.* To continue
contynued (1×),
co*n*tynued (1×)
corde, *n.* A nerve or a tendon
corde (1×),
cordys (1×)
corden, *v.* To be proper, fitting or suitable
cordyth (1×)
cordy, *a.* Stringy, fibrous
cordy (2×)
cot, *n.* An external membrane
cotys (2×),
cots (1×)
couenably, *b.* Suitably, appropriately
couenably (1×)
coueren, *v.* To cover
kou*er*ed (1×),
coueryd (1×),
couered (1×)
coueryng, *n.* The act of covering; Something that covers up or envelops
cou*er*yng (6×),
kou*er*yng (3×),
coueryng (2×),
keu*er*yng (2×)
craft, *n.* Skill
craft (1×)
cros, *n.* A cross
cros (1×)
crystallyne, *n.* The crystalline lens
crystallyne (3×),
crystalline (1×)
cupen, *v.* To bleed
Icupyd (1×)
cural, *n.* The color of red coral
cural (1×)
cut > **kutt**, *v.*
cutt > **kutt**, *v.*
cuttyng > **kuttyng**, *n.*
cyphac, *n.* The peritoneum
Cyphac (4×),
syphac (1×)

D

dedelych, *a.* Dead, deceased
dedelych (1×)
defence, *n.* The act of defending oneself
defence (1×)
defenden, *v.* To defend

defendyd (1×),
defendeth (1×),
defende (1×)
deference, *n*. Difference
deference (1×)
defyen, *v* To digest
defyed (1×)
delyng, *n*. The action of separating
delyng (1×)
demen, *v*. To judge, to pass judgement on
demyth (1×)
depe, *n*. Depth
depe (1×)
deperten, *v*. To separate
dep*er*tyd (14×),
dep*er*ted (3×),
departyd (3×),
dep*er*tyth (2×),
departed (2×)
depertyng, *n*. The action of separating
dep*er*tyng (7×),
departyng (4×)
deth, *n*. Death
deth (3×)
deuyden, *v*. To divide, separate
deuydyd (3×),
dyuyded (1×)
dew, *n*. Dew
dew (1×)
diafragma, *n*. Diaphragm
diafragma (1×)
dindim(e, *n*. The epididymis
dindimus (1×)
disesen, *v*. To be affected by an illness
disesyd (1×),
desesyd (1×)
diuerse, *a*. Of various kinds, not of one and the same kind, various
diu*er*se (2×),
dyuers (1×),
dyu*er*se (1×),
diuersse (1×),
diuers (1×)
done, *v*. To do
don*e* (6×),
done (5×),
doo (2×)
doute, *n*. Doubt
doute (1×)
down, *b*. To a lower place, down, downward
down (7×),
down*e* (4×),
downe (3×)
downward, *b*. Downward in space, from higher to lower
downward (1×)
drawen, *v*. To pull, tug, draw
draweth (1×),
Draueth (1×)
drawyng, *n*. The action of pulling
drawyng (2×),
drauyng (2×)
drede, *n*. Uncertainty
drede (1×)
drededeful, *a*. Dreadful
drededeful (3×)
dryeng, *n*. The process of drying
dryeng (1×)
drynke, *n*. Drink
drynke (2×)
duodene, *n*. Duodenum
duodenu*m* (2×)
dura mater, *n*. The outermost membranous envelope of the brain, the cranial dura mater

dura mater (4×)
duren, *v*. To continue, go on, last
dureth (1×)
dwellen, *v*. To dwell, reside
dwellyth (1×)
dyen, *v*. To die
dye (1×)
dyfferencen, *v*. To be different
dyfferenced (1×)
dyfferensyng, *n*. The action of differencing
dyfferensyng (1×)
dygestyon, *n*. The process of digesting
dygestyon (5×)
dysease, *n*. Disease
dysease (1×),
dysseyte (1×),
dysseasse (1×),
desease (1×),
disease (1×),
dyslocacion, *n*. Dislocation of a bone; also, displacement of an organ
dyslocac*i*on (1×),
dislocac*i*on (1×)
dyslocate, *a*. Dislocated
dyslocate (2×),
dyslocat (1×)
dysposicion, *n*. Constitution, complexion
dysposyc*i*on (1×)
dystroyen, *v*. To destroy
dystroyd (1×)
dyuysyon, *n*. The process of splitting or breaking up
dyuysyons (1×),
diuysyons (1×)

E

eche, *r*. Each
eche (1×)
echyng, *n*. A protuberance
echyng (1×),
echy*ng* (1×)
egge, *n*. The edge or border
egge (1×)
elbow, *n*. The elbow; An elbow
elbow (5×),
elbowe (2×)
els, *b*. Otherwise; In addition, also
els (2×)
emeroydes, *n*. The hemorrhoidal veins about the anus
emeroydes (1×)
empty, *a*. Without content
empty (1×)
ende, *n*. The end (of an action, process, condition, or period of time)
ende (35×),
endys (4×)
endelong, *b*. From end to end, longitudinally
endelong (3×)
enden, *v*. To come to an end, stop, cease
endyd (1×),
ended (1×)
engenderyng, *n*. The action of procreating
engenderyng (1×)
englych, *n*. English
englych (1×)
entren, *v*. To enter into a confinded space
entreth (5×),
entryth (3×),

entren (3×),
entre (1×)
epygloton, *n*. Epiglotis
epygloton (1×),
Epigloton (1×)
ere, *n*. One of the auricles of the heart
ere (3×),
erys (2×),
eere (1×)
escheuen, *v*. To stay away from something
eschuyd (1×),
eschew (1×)
estymacion, *n*. Comprehension, understanding
estymac*i*on (1×)
etyng, *n*. The action of eating
etyng (1×)
euen, *v*. To give birth to
euyn (1×)
euenhed, *n*. Fairness, righteousness
euenhed (1×)
euentyng, *n*. Venting, outlet
euentyng (1×)
euer, *b*. Of a continuous state or action, always
eu*er* (2×),
eu*er*e (1×)
euery, *r*. Each, every
eu*er*y (11×)
eyn > **iye**, *n*.
eyre, *n*. The gaseous substance surrounding the Earth
eyre (5×)

F

face, *n*. The human face
face (2×)
fallen, *v*. To descend, drop, fall
fallyth (2×)
fallyng, *n*. The action of falling
fallyng (1×)
fantesye, *n*. A projection of deluded or illusory imagination
fantesye (1×)
fast, *b*. Firmly
fast (3×)
fasten, *v*. To make something firm
fastned (20×),
fastened (15×),
fastnede (1×),
fastenyd (1×),
fastenede (1×)
fastnyng, *n*. The action of making something firm
fastnyng (4×),
fastenyng (2×),
fastyng (1×)
fayr, *a*. Pleasing to the sight
fayr (1×),
feyrer (1×),
farer (1×)
fayren, *v*. To make clean, purify
fayrest (1×)
felen, *v*. To feel
felt (1×),
fele (1×)
felyng, *n*. The action of feeling
felyng (8×)
femorayllys, *n*. A garment for the legs
femorayllys (1×)
fence, *n*. A fence
fence (1×)
fettyng, *n*. Bringing, conveying
fettyng (1×)
fferthermore, *b*. Furthermore
fferthermore (2×)

flank, *n*. The soft portion of the body between the ribs and the pelvis
flank (3×)
flech, *n*. Flesh
flech (6×),
flesch (2×),
flessh (1×)
fleobotomye, *n*. The operation of letting blood by opening a vein, bloodletting
fleobotomye (1×)
fleshly, *b*. In the flesh, in body, as far as the human body is concerned, physically
fleshly (1×)
flesshy, *a*. Resembling flesh, fleshlike
flesshy (2×),
fleshy (1×)
focyle, *n*. One of the two bones of the forearm
focyle (5×),
focylles (3×),
focyller (1×),
focylle (1×),
focill (1×),
focilles (1×),
focille (1×)
folden, *v*. To bent, flex
folde (2×)
foldyng, *n*. The action of flexing a limb at a joint
foldyng (1×)
folk, *n*. People
folk (1×)
folowen, *v*. To follow, come after
folowyth (1×),
folowth (1×),
folow (1×)
forcle, *n*. The pit of the stomach, the part of the abdomen in which the esophagus joins the stomach forcle (4×),
forcley (1×),
forchle (1×)
forde, *n*. A passage
forde (1×)
forhede, *n*. The forehead
forhede (9×)
former, *a*. Earlier in time
former (3×),
formor (1×),
formar (1×)
forparty, *n*. Forward or in front position
forparty (5×),
forp*er*ty (3×),
forpartye (1×)
forsayen, *v*. To say something beforehand
forsayd (5×),
forseyd (3×),
fosayd (1×)
forsoth, *b*. Truthfully
forsoth (1×),
fforsoth (1×)
forthcomen, *v*. To come forth
forthcometh (1×)
fote, *n*. A foot
fote (5×),
foote (1×),
foete (1×),
fete (1×),
feete (1×),
fote (1×)
founden, *v*. To find
founden (1×),
funde (1×),
fonden (1×)
fourme, *n*. The physical shape

of something
fourmes (2×),
fourm (1×)
ful, *a*. Full
ful (8×)
fulfyllen, *v*. To accomplish, carry out
fulfyllyd (4×),
fulfyllyth (2×),
fulfyl (1×)
fulfyllyng, *n*. The action of accomplishing
fulfyllyng (1×),
fulfyllyn*g* (1×)
fychen, *v*. To fix, fasten
fycchyd (1×)
fynger, *n*. A finger
fynger (2×),
fyngurs (2×),
fyngyr (1×),
fyngre (1×),
fyngers (1×)

G

gall, *n*. The gall bladder of man or animal
gall (1×)
gendren, *v*. To develop; To produce, bring forth
gendred (1×),
congendred (1×)
general, *a*. General
gen*er*al (1×)
gettyn, *n*. The action of acquiring, getting
gettyn (1×)
glandelouse, *a*. Containing glands, consisting of a gland or glands, glandular
glandulouse (1×)
glasin, *a*. Of humours: vitreous
glasyn (1×)
godenesse, *n*. Goodness
godenesse (1×),
godenys (1×)
gon, *v*. To go
goth (18×),
goyth (11×),
gon*e* (6×),
go (6×),
gone (4×),
goon (2×),
gon (2×),
gown*e* (1×),
gown (1×),
goo (1×),
gooth (1×),
goon*e* (1×)
goode, *a*. Good
goode (1×),
gode (1×)
goyng, *n*. The action of going
goyng (18×)
grane, *n*. Pustule, blemish on the skin
granes (2×),
graynes (1×)
gret, *a*. Large in size or quantity, big
gret (14×),
grete (12×),
grett*er* (3×),
grettest (1×),
gretter (1×),
greate (1×)
gretnesse, *n*. Size, quantity
gretnesse (2×),
gretnes (1×)
greue, *n*. A low place, a hollow
greue (1×),
graues (1×)
greyen, *v*. To become gray

greyned (2×)
greynes, *n*. Grayness
greynes (5×)
greyny, *a*. Like grain
greyny (1×)
gristyl, *n*. Cartilage, gristle
gristyl (2×),
grystle (2×),
grystyl (2×),
grystyle (1×),
grystely (1×),
gristles (1×),
grestely (1×)
growen, *v*. To grow
growyth (4×),
growen (1×),
grouyth (1×),
grouth (1×)
growyng, *n*. The action of growing
growyng (1×)
grynde, *n*. The groin
gryndes (1×)
gule, *n*. The esophagus, gullet
gule (1×)
gutt, *n*. The human intestine
gutt (8×),
gut (3×),
guttys (2×)

H

halfe, *n*. A half
halfe (15×),
halte (1×)
hamme, *n*. The part of the human leg which is behind the knee
ha*m*mes (1×)
hard, *a*. Hard
harder (3×),
harde (2×),
herde (2×),
heryd (1×),
herder (1×)
hardnesse, *n*. Hardness
hardenesse (3×),
herdnesse (1×),
hardenysse (1×),
hardnesse (1×)
harme, *n*. Pain
harme (1×)
haunche, *n*. The hip
hanche (2×),
haunche (2×),
haunches (2×),
henches (1×),
hanches (1×),
hanch (1×),
haunches (1×),
hauches (1×)
hauyng, *n*. The action of having
hauyng (8×),
hauy*ng* (2×),
hauyn (1×)
haven, *v*. To have
hath (21×),
haue (8×),
hauen (1×),
hadde (1×)
hede, *n*. The head
hede (29×),
hed (4×),
hedys (2×),
hedde (1×),
hede (1×)
helen, *v*. To cure
helyd (3×)
helpen, *v*. To help
helpyth (2×),
helpe (1×)
helpyng, *n*. The action of helping

helpyng (2×)
hem, *r*. Them
hem (22×)
hemselfe, *r*. Themselves
hemselfe (1×)
her, *b*. Here
her (12×),
here (5×)
hery, *a*. Hairy
hery (1×)
heryng, *n*. The action of hearing
heryng (2×)
herte, *n*. The heart
herte (7×),
hert (7×)
heryd > **hard**, *a*.
hete, *n*. Heat
hete (1×)
heten, *v*. To heat
hete (1×)
heyle, *n*. The heel
heyle (2×),
heele (1×)
holden, *v*. To enclose, contain
holdyth (2×),
holdeth (1×)
holdyng, *n*. The action of enclosing, containing
holdyng (4×)
hole, *n*. A perforation, hole
hole (21×),
holys (9×),
holes (3×)
holowe, *a*. Having a cavity inside, hollow; of a surgical instrument: tubular, hollow
holowe (4×),
holow (4×),
holough (1×)
holowenesse, *n*. The condition of being hollow
holownesse (10×),
holownese (4×),
holownes (2×),
holownys (1×),
holounesse (1×)
honde, *n*. The human hand
honde (9×),
hondys (1×)
hoop, *n*. Something circular
hoopes (1×)
horne, *n*. One of the horns on the head of certain quadrupeds
horne (1×)
how, *b*. In what manner or way
how (7×),
howe (1×)
houen, *v*. To wait in readiness or expectation
houeth (1×)
humore, *n*. One of the four fluids which form and nourish the body
humore (3×),
humoure (2×),
humors (2×),
humours (1×),
humor (1×)
hurt, *n*. A wound, an injury
hurt (1×)
hurten, *v*. To injure
hurteth (1×),
hurt (1×)
hurtyng, *n*. The act of hurting
hurtyng (6×)
hyden, *v*. To hide
hyd (2×),
hydde (1×)
hynder, *a*. Lying or situated in the rear, back
hynd*er* (10×),

hynder (2×),
hynde (1×)
hyppen, *v.* Of persons, animals, things: to leap
hyppyth (1×)
hyr, *d.* With reference to female persons: her
hyr (1×)

I

iaw, *n.* Jaw
Iawys (1×),
Iawes (1×),
Iawe (1×)
inwarde, *b.* To or toward the inside of a building, space, opening
inwarde (2×)
inbylen, *v.* To enter
inbylen (1×)
innere, *a.* Inner, inside
Inner (1×)
innermer, *a.* Inner, interior
inn*er*mer (1×)
insencyble, *a.* Incapable of being felt or perceived by the senses
insencyble (1×)
instape, *n.* The arch of the foot
instape (1×)
instrument, *n.* A part or organ of the body performing a function
instrument (2×),
i*n*strument (1×)
ioynen, *v.* To join
Ioyned (6×),
ioned (2×),
yoned (1×),
ioynen (1×)
ioynyng, *n.* The action of joining
ioynyng (1×),
Ionyng (1×)
ioynt, *n.* A joint between bones; either the place where the bones meet or the structure holding them together
ioyntes (3×),
Ioyntys (1×),
ioynters (1×),
Ioynte (1×)
ioynture, *n.* A joint between bones
Ioynture (1×)
iþhen, *n.* A swelling filled with phlegmatic humors, edema
Iþhen (1×),
Iþeum (1×)
iye, *n.* The eye
Iye (5×),
Iyen (3×),
eyn (1×)

K

kepen, *v.* To keep
kepyth (1×),
kepte (1×),
kept (1×)
kepyng, *n.* The action of keeping
kepyng (1×)
keueryng > **coueryng**, *n.*
kne, *n.* The human knee
kne (6×)
knotte, *n.* A joint in the body
knotte (1×),
knote (1×)
knotty, *a.* Tied full of knots; Knotted
knotty (2×),
knottly (1×),

knotly (1×)
knowen, *v*. To know
knowne (1×),
knowene (1×),
knowen (1×),
know (1×),
knawe (1×)
knowyng, *n*. The action of knowing
knowyng (1×)
knowlege, *n*. Knowledge
knowlege (1×)
knytten, *v*. To fasten by a knot
knyt (5×),
knytte (2×),
knit (2×)
knyttyng, *n*. The action of fastening by a knot
knyttyng (5×),
knyttyn (1×),
knettyng (1×)
kouered > **coueren**, *v*.
kouerуng > **coueryng**, *n*.
kut, *v*. To cut
kut (3×),
kuþt (2×),
cutt (2×),
cut (1×)
kuttyng, *n*. The act of cutting;
kuttyng (12×),
cuttyng (6×)
kydnere, *n*. The kidney of a person or an animal
kydnerys (1×)
kynde, *n*. The natural constitution of a person; man's physical nature
kynde (13×),
kynd (2×)
kyndely, *b*. According to the regular course of nature, as a result of natural causes
kyndely (4×),
kyndly (1×)

L

lappen, *v*. To wrap
lapped (1×)
large, *a*. Ample in quantity
large (3×)
last, *a*. Last, final
last (7×)
lasten, *v*. To last
lastyth (1×)
latter, *a*. Later
latter (1×)
leche, *n*. A physician or surgeon
leche (6×),
lechys (1×)
ledden, *v*. To conduct
ledde (1×)
left, *a*. Left
left (10×),
lyfte (1×),
lefte (1×)
legg, *n*. The human leg
legg (9×)
length, *n*. Linear extent or distance
length (15×),
lengthe (1×),
length*e* (1×),
lenght (1×)
lepen, *v*. To jump
lepe (1×)
lesse, *a*. Less
lesse (6×),
lasse (1×)
lest, *b*. At last, finally
lest (1×)
letten, *v*. To allow
lettyd (2×),

lettyth (1×),
lette (1×)
letyng, *n.* The action of allowing
letyng (3×),
lettyr (1×),
lettyng (1×)
long, *a.* Longer; Long
long (3×),
lenger (2×)
longaon, *n.* The rectum
longaon (1×),
langaon (1×)
loosse, *a.* Free to move about
loosse (1×)
lowe, *a.* Lower in space; Lowest in space
lower (2×),
lowest (1×),
lauer (1×)
lowen, *v.* To make something low
loweth (1×)
luddok, *n.* The loin
luddokys (1×)
luine, *n.* A length of cord
luine (1×)
lung, *n.* The human lung
lung (5×),
longes (1×),
long (1×)
lyen, *v.* Dwell, abide
lyeth (1×),
lyen (1×)
lyffe, *n.* Life, lifetime
lyffe (2×)
lyften, *v.* To raise something from the ground or other surface
lyftyth (1×)
lyggyng, *n.* Being in a reclining posture; An act of sexual intercourse
lyggyng (3×)
lyght, *a.* Light in weight
lyght (2×)
lyghtly, *b.* With little effort, without difficulty
lyghtly (3×)
lyke, *p.* Similar to
lyke (8×)
lyknesse, *n.* Similarity
lyknesse (1×),
lyknes (1×)
lynk, *n.* A vertebra
lynkys (2×),
lynk (1×),
linkys (1×)
lytyl, *a.* Little
lytyl (9×)
lytyl fyggyr. The little finger
lytyl fyggyr (1×)
lyuer, *n.* The human liver
lyuer (8×),
lyu*er* (5×),
lyuor (1×)

M

maken, *v.* To make
made (39×),
mayd (28×),
makyth (6×),
make (2×),
mayde (2×),
ymade (1×),
maken (1×),
Imayd (1×)
makyng, *n.* The action of making
makyng (17×),
maky*ng* (2×)
malyncolyose, *a.* Consisting of or mixed with black bile

malyncolyose (1×),
malycolyus (1×)
man, *n*. Man; A person, a man or woman
man (5×),
man*e* (2×),
mannys (1×)
maner, *n*. A kind, sort, variety
man*er* (25×),
maner (7×)
many, *d*. Many
many (4×)
marow, *n*. Human marrow, the tissues containing fat and blood cells in the cavities of bones
marow (4×),
marown (1×),
marowe (1×)
marowy, *a*. Full of marrow
marowy (1×),
marowshy (1×),
marowhy (1×)
maryce, *n*. The uterus
maryce (2×),
marice (1×)
matrix, *n*. A blood vessel; The uterus, womb
matrix (3×),
matrix (1×)
mayen, *v*. To be able, be capable
may (10×),
mowen*e* (1×),
mow (1×),
mayst (1×),
maye (1×)
meddelyng, *n*. The action of mixing or mingling
meddelyng (2×),
medlyng (1×),
medelyng (1×),
medelyngys (1×)
medlen, *v*. To blend, mix;
medlyde (1×),
medled (1×),
medelyd (1×),
meddelyd (1×)
medycyne, *n*. A medicine
medycyne (1×)
medyll, *n*. Middle region in space, the middle part of a thing
medyll (2×),
middyl (1×),
medyl (1×)
membre, *n*. A bodily limb or organ
membres (13×),
membre (9×),
membrys (4×),
membros (3×),
membris (2×),
membres (2×),
me*m*bre (1×),
membrox (1×)
mery, *n*. The esophagus
mery (1×)
meryngg > **myryng**, *n*.
mesure, *n*. Measure
mesure (1×)
mete, *n*. Food, nourishment, sustenance
mete (3×)
meuen, *v*. To move
mouyd (3×),
moueth (2×),
mouen (2×),
mouyth (1×),
mouyn (1×),
moue (1×)
moder, *n*. Mother
moder (1×)

more, *a*. Large in size or spatial extent
more (24×)
mouth, *n*. The human mouth
mouth (4×),
mouthe (1×),
mouthe (1×)
mouyng, *n*. The action of moving
mouyng (22×),
mouy*ng* (4×),
mouynng (1×),
ammouyng (1×)
moyst, *a*. Most
moyst (4×),
most (3×)
moyste, *a*. Moist, wet
moyst*er* (1×)
moysten, *v*. As modal auxiliary: must
moyste (1×),
moyst (1×)
moystnesse, *n*. Moisture, wetness
moystnesse (1×)
much, *a*. Much
much (2×),
moche (1×)
multyplyen, *v*. To become greater in size
multyplyeth (1×)
multytude, *n*. A large number of persons or things
multytude (1×)
muscle, *n*. A muscle
muscles (20×),
muskles (4×),
musclys (3×),
musches (1×),
muskelys (1×),
muskeles (1×)
myddys, *n*. The middle part of something
myddys (10×),
mydreste (1×)
mydryfe, *n*. The diaphragm
mydryfe (3×),
myddrefe (2×),
myddryfe (2×),
myddryf (1×),
myddref (1×)
mydward, *n*. The middle of a thing
mydward (1×)
myghten, *v*. Might
myght (2×)
myghtty, *a*. Mighty, powerful
myghtty (1×)
myghtyly, *b*. Powerfully
myghtyly (1×)
mylte, *n*. The spleen
mylte (1×),
mylt (1×),
melt (1×)
mynde, *n*. The human mind
mynde (2×)
myryng, *n*. Union, joining
myryng (2×),
meryngg (1×)

N

namely, *b*. Especially
namely (1×)
narrow, *a*. Narrow
narrow (1×)
naturally, *b*. In accord with natural law
naturally (1×)
nature, *n*. Nature as governed by law
nature (2×)
nauicula, *n*. The navicular bone

of the foot
nauicula (2×)
nauyl, *n.* The human navel, umbilicus
nauyl (2×)
necessarye, *a.* Necessary
necessarye (1×),
necessayre (1×),
necessarie (1×)
nede, *n.* Need
nede (2×)
nedeful, *a.* Needful
nedeful (6×)
neden, *v.* To need
nedyth (4×),
nede (2×)
neke, *n.* The neck
neke (20×),
nek (5×)
nere, *a.* Near
nere (1×)
neshe, *a.* Physically soft in texture
neshe (1×),
nesche (1×)
nether, *a.* Lower; under
nether (14×),
neþ*er* (11×),
neþer (1×)
neþerles, *b.* Nevertheless
neþ*er*les (1×)
neþermer, *b.* Lower
neþ*er*mer (1×)
nette, *n.* Something constructed like a net
nette (2×),
nett (1×)
neuermore, *b.* No longer
neu*er*more (1×)
neuerthelese, *b.* Nevertheless
neuerthelese (1×),
neu*er*thelasse (1×)
new, *a.* New
new (1×)
nexte, *a.* Next
nexte (6×)
neyþer, *c.* Neither
neyþ*er* (1×)
nigra, *n.* The median vein
nigra (1×),
Nigra (1×)
noble, *a.* Of high rank in the hierarchy of created things
noble (3×)
nor, *c.* Nor
nor (2×)
noryshen, *v.* To supply (sb.) with food or drink, feed
noryshyd (4×),
noryshyth (2×),
norysch (1×)
noryshyng, *n.* The action of feeding, furnishing with nutriment
noryshyng (5×)
nose, *n.* The nose as part of the human body
nose (11×)
nost, *a.* Having nostrils of a certain kind
nost (1×)
nothyng, *r.* Nothing
nothyng (2×)
nought, *b.* Not
nought (2×)
now, *b.* Now, at the present time
Now (2×)
nowmbre, *n.* A number
nowmbre (2×),
noumbre (1×)
nowmbren, *v.* To count
nowmbred (3×),

noysaunce (1×)
nucha, *n.* The spinal cord
nucha (11×)
nygh, *b.* Near in space
nygh (1×),
nyz (1×),
nyȝ (1×)
nyz > **nygh**, *b.*
nyȝ > **nygh**, *b.*

O

oft, *b.* Repeatedly, again and again, many times, frequently
oft (1×)
only, *b.* Solely, only, exclusively
only (3×),
ony (1×)
onythyng, *n.* Anything
onythy*ng* (1×)
open, *n.* Aperture
open (1×),
opyn*e* (1×),
opyn (1×),
open, *a.* Open
open (6×),
opyn (2×),
opyn*e* (1×),
opon (1×),
open*e* (1×),
opon*e* (1×)
oponen, *v.* To open
oponeth (1×),
oponed (1×)
opynly, *b.* Openly
opynly (10×),
openly (5×),
oponly (2×)
ordeinyng, *n.* The action of preparing, making ready
ordeinyng (1×)
ordeynen, *v.* To prepare, make ready
ordeyned (12×),
ordenyd (10×),
ordeynyd (3×),
ordenyde (1×),
ordened (1×),
ordeynd (1×)
ordure, *n.* Filth, dirt
ordure (1×)
os, *n.* A bone
os (3×)
other, *r.* The second of two persons or things
other (13×),
oþ*er* (9×),
oþer (1×),
oder (1×)
otherwhyle, *b.* Sometimes, occasionally
otherwhyle (1×),
oþ*er*whyle (1×)
ouer, *b.* Over
ouer (3×),
ou*er*e (1×),
ou*er* (1×)
ouerthwert, *b.* Across
ouerthwert (1×),
ou*er*thwerte (1×)
outcastyng, *n.* The action of ejecting or casting forth or away also
outcastyng (1×)
outegoyng, *n.* A going out
outegoyng (1×)
outtaken, *v.* To take something out, to remove
outtakyth (1×)
outward, *b.* Away from or out of a place or position

outward (1×),
outeward (1×)
owen, *v.* To owe
owen (8×),
own*e* (1×),
owen*e* (1×),
owe (1×)

P

pacyent, *n.* One who suffers from an injury or a disease, a sick man
pacyent (1×)
palet, *n.* Head
palet (4×)
panne, *n.* A garment, esp. a cloak or mantle
panne (2×)
pannicle, *n.* A membrane covering an organ, muscle, bone, etc.
pannicle (8×),
pannycle (4×),
pannycles (1×),
pannicles (1×)
parten, *v.* To divide
partyd (1×)
passen, *v.* To move, advance
passyth (5×),
passeth (1×),
passen (1×),
passe (1×)
passyng, *n.* The action of passing
passyng (5×)
paxillus, *n.* The sphenoid bone
paxillus (3×),
paxillare (1×),
Paxillus (1×),
Paxillare (1×)
payre, *n.* A pair, a set of two
payre (14×),
payer (4×),
paire (3×),
payr (3×)
pele, *n.* Skin
pele (1×)
perfyte, *a.* Ideal, supreme, flawless
p*er*fyte (2×)
persen, *v.* To cause an incision, penetrate
persyd (2×)
persen, *n.* A person
p*er*sens (1×)
persyng, *n.* The action of percing or penetrating
p*er*syng (4×),
p*er*chyng (1×)
perty, *n.* Anatomical part
p*er*tye (18×),
p*er*ty (16×),
party (14×),
partye (12×),
p*er*tyes (11×),
partyes (6×),
partys (5×),
parties (4×),
partie (3×),
p*er*tie (2×),
perty (1×),
part*es* (1×),
p*er*tys (1×),
part (1×),
p*er*tye (1×)
pia mater, *n.* The soft inner membrane covering the brain, the pia mater
pia mater (6×)
pix, *n.* A socket
pixis (1×)
place, *n.* Place, locality, spot

place (30×),
places (15×),
placys (2×),
placis (1×),
places (1×)
plant, *n.* A plant
plantys (1×)
playstre, *n.* A plaster, poultice or compress made from herbs
playstres (1×)
possible, *a.* Possible
possible (1×)
poure, *n* An internal channel for bodily fluids such as blood, urine, etc
pouris (1×)
power, *n.* The ability of a person
power (1×)
prencipallych, *b.* Mostly, mainly, primarilly
p*re*ncypallych (1×)
preuy, *a.* Evident, manifest
preuy (4×),
preuye (1×)
preuyly, *b.* Manifestly, clearly
preuyly (1×)
profitable, *a.* Advantageous, profitable
p*ro*fitable (3×)
proprely, *b.* Properly
proprely (2×),
proprelyche (1×)
prykken, *v.* To pierce, stab
prykkyd (1×)
pourgen, *v.* To purify or cleanse
pourged (1×)
pupilla, *n.* The pupil of the eye
pupilla (1×),
Pupilla (1×)
purchasen, *v.* To acquire
purchesyth (1×),
purchesyd (1×)
purpurea, *n.* The median cubit vein
purpurea (1×),
Purpurea (1×)
putten, *v.* To put
put (5×),
puttyth (1×)
puttyng, *n.* The action of putting
puttyng (2×)
pyle, *n* Human skin
pyles (1×)
pyngthen, *v.* To pierce, stab
pyngth*e* (1×)
pyntyl, *n.* A human penis
pyntyl (1×)
pype, *n.* A tube, duct or vessel of the human body
pype (5×),
pypys (4×),
pypes (1×),
pipe (1×),
pipes (1×)
pyt, *n.* A natural depression or hollow in some part of the body
pyt (2×),
pytt (1×)

Q

quantyte, *n.* Quantity, amount
qua*n*tyte (1×)
quyken, *v.* To sustain (a part of the body) in life
quykyn*e* (1×),
quykkyn (1×),
quykkyd (1×),
quykkenyd (1×),
quyken*e* (1×)

R

raffe, *n* A garment, esp. a cloak or mantle
raffe (1×)
rall, *n* The kneecap, patella
rall (1×)
rasche, *n.* The cuneiform and cuboid bones in the foot collectively
rasche (1×)
reme, *n* A skin
reme (1×)
remedy, *n.* A medical remedy
remedy (1×)
remnaunt, *n.* The rest
remnaunt (1×)
remouen, *v.* To take away, remove
remouyd (1×)
remouyng, *n.* The action of taking away, removing
remouyng (1×),
re*m*meuyng (1×)
rennen, *v.* To run
rennyth (1×),
renneth (1×)
replete, *a.* Filled, covered
replete (1×)
resauyen, *v.* To ingest
resauyn*e* (1×),
resauyn (1×)
resayen, *v.* To endow
resayneth (1×)
restyng, *n.* The action of resting
restyng (1×)
restoren, *v.* To restore
restoryd (2×)
restoryng, *n.* The action of restoring
restoryng (3×)
reyne, *n.* A human or an animal kidney
reynes (11×)
ribbe, *n* A human rib
rybbys (10×),
rybbys (1×)
rigge, *n.* The back of a person
rygges (1×),
ryge (1×)
roll, *n.* A scroll
roll (5×),
rol (1×)
ronnyng, *n.* The action of running
ronny*n*g (1×),
re*n*nyng (1×)
rote, *n.* Root
rote (3×),
rotys (2×),
rote (1×)
rownde, *a.* Spherical in shape
rownde (5×),
rounde (4×),
rou*n*de (1×)
rowndnesse, *n.* The quality of being spherical
rowndnesse (1×),
rowndenesse (1×)
rubbyng, *n.* The action of rubbing
rubbyng (3×)
ryght, *a.* Right, as opposed to left
ryght (11×)
ryght, *b.* As intensifier, very extremely
ryght (2×)
righttyng, *n.* Making correct, putting right
ryghttyng (1×)

S

same, *a*. Identical; equal
same (10×)
saue, *p*. Except, except for
saue (2×),
safe (1×)
sawe, *n* A saw
sawe (3×)
schape, *n*. Form, shape; The pubic bone
schape (16×),
schap (9×),
shape (2×),
schaape (1×),
schape (1×)
schare, *n*. The pubic bone
share (4×),
shares (1×),
schare (1×)
schewen, *v*. To show
schewyth (5×),
shewyth (3×),
schewyde (1×),
schewyd (1×),
schewen (1×),
schew (1×)
schewyng, *n*. The action of showing
schewyng (1×)
schort, *a*. Short
schort (2×),
schorte (1×)
schortnesse, *n*. Shortness in extent or distance
schortnesse (1×)
schuldre, *n*. The shoulder of a person
schuldre (12×),
schulder (5×),
schulders (3×),
scholdre (2×),
shuldre (1×)
schulen, *va*. As modal auxiliary: shall
schal (6×),
schuld (4×),
scholde (4×),
schalt (3×),
schulde (2×),
shold (1×),
shal (1×),
schul (1×),
schold (1×)
sed, *n*. Seed
sede (1×)
sell, *n*. A harness
sell (1×)
semely, *b*. In a proper manner, fittingly,correctly
semely (1×)
semen, *v* Seem; also, seem to be
semyth (10×),
semeþe (1×)
sen, *v*. To see
sygthty (1×),
sought (1×),
seen*e* (1×),
se (1×)
sencyble, *a*. Capable of being sensed or felt, perceptible
sencyble (4×)
seke, *a*. A sick person, patient
seke (1×)
sekenes, *n*. Illness, sickness
sekenes (2×),
sekenesse (2×),
syknesse (1×)
sekernys, *n*. Freedom from danger, physical safety, security
sekernys (1×)

senden, *v*. To send
sendyth (2×)
sesen, *v*. To seize
sesyth (1×)
seten, *v*. To set
set (7×),
sett (4×),
vnset (1×)
seyen, *v*. To say
seyd (5×),
sey (4×),
seyn (1×),
seyden (1×),
say (1×),
saye (1×),
seyde (1×)
sharpnes, *n*. Sharpness
sharpnes (1×)
sheten, *v*. To close
shet (1×)
shynnen, *v*. To shine
shynne (1×)
sklyroticus, *n*. The sclerotic coat of the eye, the white of the eye
sklyroticus (1×)
skul, *n*. The skull
skul (13×), skull (1×)
skyn, *n*. Skin
skyn (35×),
skynne (18×),
skynnys (5×),
skyn*e* (3×),
skynnes (2×),
skyne (2×),
skyn*e* (1×)
skyrotus, *n*. The white of the eye, the sclera
skyrotus (1×)
slepe, *n*. The state of sleep
slepe (1×),
sclepe (1×)
slyme, *n*. A slimy substance; mucus
slyme (1×)
smal, *a*. Small
smale (9×),
smaller (1×),
smalle (1×)
smellyng, *n*. The sense of smelling
smellyng (2×)
so, *b*. So
so (25×)
sodeyn, *b*. Suddenly, unexpectedly
sodeyn (1×)
soft, *a*. Softer; Soft
soft (2×),
softe (2×),
soyft (1×),
softer (1×)
softnesse, *n*. Softness
softnesse (2×),
softnes (1×)
sore, *n*. A sore
sore (2×)
sotel, *a*. Secure, firm
sotel (1×)
soth, *a*. Real, actual
soth (1×)
sothe, *b*. Truthfully, in truth
sothe (1×),
Soth (1×),
soþ*er* (1×)
sothly, *b*. Truly
Sothly (5×)
sought > **sen**, *v*.
sowle, *n*. The mind, intellect; also, the imagination
sowle (2×),
soule (2×)

sownyng, *n.* A state of temporary unconsciousness
sownyng (1×)
spawd, *n.* The shoulder blade of a person
spawd (1×)
specyally, *b.* Especially
specyally (6×)
spedy, *a.* Beneficial, helpful
spedy (1×)
spekyng, *n.* The action of speaking
spekyng (1×)
sperler, *n.* The calf of the leg
sperler (1×)
sperme, *n.* The male seminal fluid
sperme (8×),
sperm (2×),
sperma (1×)
splene, *n.* The spleen of a human
splene (3×)
spondel, *n.* A vertebra
spondels (20×),
spondyls (5×),
spondyl (3×),
spondyle (3×),
spondelys (3×),
spondele (3×),
spondel (3×),
spondys (1×),
spondell (1×)
spreden, *v.* To lay, spread
spred (4×),
spredyth (2×),
sprede (2×)
spyritual, *a.* Immaterial, consisting solely of intellect
spyritual (1×),
spiritual (1×)
spyryt, *n.* Spirit, soul
spyrytys (2×),
spyrites (1×),
spryte (1×)
standyng, *n.* The action of standing
standyng (1×)
stape, *n.* The sole of the foot
stape (1×)
stede, *n.* A particular part of space
stede (2×),
steedys (1×)
steren, *v.* To control
steryth (1×)
steryng, *n.* The action of controlling
steryng (1×)
stomake, *n.* Stomach
stomake (12×),
stomak (1×),
stomake (1×)
ston, *n.* A stone
stonys (1×),
stony (1×),
stone (1×)
stoppen, *v.* To stop
stoppyd (2×)
strayte, *a.* Narrow, not wide
strayte (2×),
strayt (1×),
strate (1×)
strecche, *n.* A way
strecche (1×)
strechen, *v.* To stretch
strechyth (1×),
strechyd (1×)
strechyng, *n.* The action of stretching
strechyng (2×)
strength, *n.* Physical strength

strength (1×)
stryng (1×),
strong, *a*. Stronger; Strong
strenger (3×),
strong (2×),
stronge (2×),
streng*er* (2×),
stryngy (1×),
strenges (1×),
streng (1×)
strongly, *b*. With physical strength or force
strongly (1×)
styptyk, *n*. Sourish
styptyke (1×)
substance, *n*. A type or kind of thing
substance (7×)
subtyle, *n*. A portion of small intestine
subtyle (1×)
such, *d*. Such
such (4×),
suche (3×)
sundry, *a*. Different in nature
sundry (1×)
superfluyte, *n*. A bodily part formed from an excess of humors
sup*er*fluytes (3×),
sup*er*fluyte (2×)
supportyng, *n*. The action of supporting
supportyng (1×)
susteynen, *v*. To support
susteyn (1×)
suttyle, *a*. Manifest, clear, evident; easily understandable; unambiguous
suttylte (1×)
swellyng, *n*. Morbid enlargement of a body or a bodily part because of disease
swellyng (2×)
syde, *n*. Side
syde (10×),
sydes (1×)
syght, *n*. The faculty of sight
syght (2×)
sygthty > **sen**, *v*.
sykyr, *a*. Free from danger, not at risk
sykyr (1×),
syker (1×)
sympely, *b*. Sincerely, honestly
sympely (1×)
synew, *n*. A fibrous cord connecting a muscle to a bone or other bodily part, a tendon
synews (29×),
synewes (8×),
synew (8×),
senues (6×),
synues (6×),
senews (3×),
synewe (2×),
synue (1×),
syneus (1×),
synnues (1×),
senewes (1×),
syneuys (1×),
seneuys (1×)
synewy, *a*. Tendinous
synewy (2×),
synowy (1×)
synuesse, *n*. The quality of having many tendons
synuesse (2×),
synuysse (1×)

syphac > **cyphac**, *n.*

T

taken, *v.* To take; To take, have
taken (3×),
take (3×),
Itaken*e* (1×)
takyng, *n.* The action of taking
takyng (2×)
tayle, *n.* The lowest portion of the spinal column
tayle (1×)
tellen, *v.* To tell
tell (1×),
tolde (2×)
tery, *a.* Producing tears
tery (1×),
Tery (1×)
tet, *n.* A nipple, teat
tetys (1×)
tethe > **toth**, *n.*
thene, *b.* Then; There, therein
thene (1×),
þen*e* (1×),
þen (1×)
ther, *b.* There
þ*er* (26×),
ther (12×),
þer (7×),
There (3×),
þ*er*er (1×)
therfor, *b.* Therefore
Therfor (3×),
þ*er*for (3×),
Therfore (3×),
þ*er*fore (2×)
þerto, *b.* Thereto
þ*er*to (2×)
þeuen, *v.* To serve
þeuen (3×),
þeueth (2×),
þeue (1×)
þilk, *a.* Referring to a specific member of a class and used anaphorically
þilk (2×),
þilke (2×),
thylke (1×),
thylk (1×)
thorax, *n.* The chest cavity
thorax (4×),
torax (1×)
þorough, *p.* Through
þorough (1×)
þough, *c.* In spite of the fact that
þough (1×)
thought, *n.* Thought, the product of mental activity
thought (1×)
thoume, *n.* The thumb
thombe (4×),
thome (1×)
thriven, *v.* To grow, increase, mature
thryueth (1×)
throte, *n.* Throat
throte (12×),
þrote (2×),
throote (1×)
thyȝ, *n.* The upper leg of a human being, thigh
thyȝ (2×),
thyz (1×),
thyes (1×),
þyes (1×)
thykke, *a.* Thick, dense; Thicker
thykker (1×),
thykke (1×)
thyknesse, *n.* Thickness
thyknesse (1×)
thynne, *a.* Thin

thynne (2×),
þin (1×)
thyng, *n.* Thing
thyngs (2×),
thyng (1×),
thy*n*g (1×),
þyngys (1×)
time, *n.* Time
tyme (10×)
to, *n*. A toe of the human foot
toes (4×),
tooes (2×),
too (2×),
toes (1×)
tochen, *v*. To touch
tocheth (1×)
toder, *r*. The second of two things, the latter
tod*er* (1×)
togydder, *b*. Together
togydd*er* (9×),
togyddyr (5×),
togydder (5×),
togydd*er*s (1×),
togedder (1×),
togeder (1×)
token, *n*. Proof, evidence, an observable characteristic
token (1×)
tong, *n.* The tongue of a person
tong (4×),
tonge (1×)
toth, *n* A tooth of a human
tethe (3×)
toward, *p*. Toward, in the direction of
toward (5×),
towarde (4×)
trachea, *n*. The tube connecting the larynx and the bronchi, the windpipe
Trachea (1×)
tre, *n*. An object made of wool
tre (1×)
trew, *a*. Truthful, veracious; not given to deceiving, misleading, or dissembling
trewer (1×)
turnen, *v*. To change position or orientation
yturned (1×),
turnyth (1×),
turned (1×)
tyen, *v*. To fasten
tyed (2×),
tyde (1×)
tyl, *b*. Until
tyl (2×),
tyll (1×)

U

uentrycle, *n*. A chamber or cavity within a bodily organ
uentrycle (2×),
ventrycles (1×),
wentrycle (1×),
wentryclys (1×),
ventrycle (1×),
uentrycles (1×),
uentricle (1×),
uentricles (1×),
uent*r*icles (1×),
uentryclys (1×)

V

veile, *n.* A veil
veile (1×)
ventusyng, *n*. The application of suction to a body part without the promotion of bleeding

ventusyng (1×)

vernayle, *n.* Of a red or bright color, vermilion
vernaylle (1×),
werneyllys (1×),
vernayle (1×)

vertue, *n.* Physical ability or skill
vertue (4×),
wertu (1×),
vertues (1×),
uertues (1×),
vertu (1×),
vertues (1×)

vessyl, *n.* A vein or an artery
vessyl (3×),
vesyl (1×),
vessyls (1×),
vassyl (1×),
vessels (1×)

veyne, *n.* A blood vessel, vein
veynes (33×),
veyne (16×),
veynys (12×),
vayne (4×),
vaynys (3×),
weynes (2×),
veyn (2×),
venys (2×),
vaynes (2×),
wanys (1×),
vanyes (1×),
vayn (1×),
veynnys (1×),
venyes (1×),
vanis (1×),
veyny (1×)

vnable, *a.* Unable, incapable
vnable (1×)

vncurable, *a.* Not amenable to medical treatment, incurable
vncurable (1×)

vnder, *p.* Under
vnder (14×),
vnd*er* (12×),
Under (2×),
vndyr (1×)

vnderberen, *v.* To support
vnderberyth (1×)

vnderstonden, *v.* To understand
vnderstond (1×)

vnkindly, *b.* In a manner contrary to the regular course of nature
unkyndly (1×)

vnknouen, *v.* To be or remain undiscovered
vnknown (2×)

vnsencyble, *a.* Incapable of feeling
vnsencyble (1×)

vnset > **seten**, *v.*

vpper, *a.* Upper
vpper (29×)

vppermore, *a.* Higher, farther up toward the top
vppermore (2×),
vppermor (1×)

vppermost, *a.* Pertaining or proper to the top part or highest region of something
vppermost (1×)

vpward, *b.* In the upper part, at the top, above
vpward (2×)

voice, *n.* The voice of a human being
voices (2×),
voyces (1×),
voice (1×)

vouen, *v.* To support
wouyn (2×),

wouen (1×)
voyde, *a*. Without content, empty
voyde (1×)
vriclides, *n*. The ducts that transport urine from the kidneys to the bladder, the ureters
vriclides (1×)
vryn, *n*. Urine
vryn (3×),
vryne (1×)
vtter, *a*. Outer
vtter (5×),
vtt*er* (2×),
vntt*er* (1×)
vttermore, *a*. Outer, exterior
vttermore (5×),
utt*er*more (2×),
vtt*er*mare (1×),
vtt*er*more (1×)
vttermost, *a*. Outermost
vttermost (1×)

W

waxen, *v*. To develop physically
waxeth (2×),
vaxeth (1×)
waxyng, *n*. The action of developing physically
waxyng (1×)
wel, *b*. Rightly; well
wele (5×),
wel (2×),
weyl (1×),
well (1×),
welle (1×)
wenden, *v*. To make one´s way, move
ueth (4×)
wentrycle > **uentrycle**, *n*.
wentryclys > **uentrycle**, *n*.
wenyng, *n*. The action of gaining control
wenyng (1×)
werneyllys > **vernayle**, *n*.
werchyng > **worchyng**, *n*.
weten, *v*. To dampen or moisten
wete (6×),
wet (1×),
vete (1×),
uete (1×)
wetyn, *n*. The action of dampening or moistening
wetyn (1×)
wey, *n*. Way
wey (2×),
weyes (2×),
weye (1×)
whan, *c*. When
whan (20×),
whanne (1×),
whan*e* (1×)
what, *a*. What, what sort or sorts of
what (4×)
whateuer, *r*. Referring to things
whateu*er* (1×)
where, *b*. Where
where (2×)
wherfore, *c*. With reference to an action
wherfore (4×)
whole, *a*. Whole
whole (1×)
whom, *r*. Whom
whom (2×),
whome (1×)
whose, *r*. Whose
whose (6×)
why, *c*. Why
why (7×)

whych, *r*. Which
whych (158×),
wych (7×),
which (2×),
whyche (1×)
whyrlbone, *n*. The hip or hip joint
whyrlbone (3×),
whylrbone (1×)
whyte, *a*. White
whyte (4×),
whytte (1×)
wombe, *n*. The human uterus, womb; A cavitiy within the human body
wombe (6×),
wombys (1×)
woman, *n*. A woman
woman (3×),
women (1×),
woma*n* (1×)
worchyng, *n*. The action of working
worchyng (10×),
wyrchyng (1×),
worchy*ng* (1×),
werchyng (1×)
wond, *n*. An injury
wondys (1×),
wonde (1×)
wonden, *v*. To penetrate
wonden (2×),
wounded (1×)
wrynkle, *n*. A curving passageway
wryncles (2×)
wryst, *n*. The wrist
wryst (1×)
wyde, *a*. Wide
wyde (1×)
wylful, *a*. Eager to do something
wylful (1×)
wylfully, *b*. In a headstrong or stubborn manner
wylfully (1×)
wylen, *v*. To have desire for something; want something to occur, wishl; As modal auxiliary: will
wyl (2×),
wolde (1×)
wynde, *n*. Gas or air within the body
vynde (1×)
wyrchyng > **worchyng**, *n*.
wyssen, *v*. To guide on a route
wyst (1×)
wytt, *n*. Wit
wytt (3×),
wyt (1×)

Y

ȝarde, *n*. A penis
ȝarde (6×),
ȝerde (2×),
ȝerd (1×),
yarde (1×)
ȝelding, *n*. The act of yielding a substance
ȝeldyng (1×)
ȝerbus, *n*. The first part of the large intestine: the caecum.
ȝerbus (1x)
ȝi, *a*. High, lofty, big, tall; High
ȝi (2×)
yllyon, *n*. The portion of the small intestine which opens into the caecum
yllyon (1×)
ymagynacion, *n*. Imagination
ymagynac*i*on (1×)
ynche, *n*. A measure of length

equal to the twelfth part of a foot
ynches (1×)

yrene, *n.* Iron
yren*e* (1×)

ysophagus, *n.* The esophagus
ysophagus (1×)

ȝyfen, *v.* To give
ȝyf (1×)

CHAPTER FIVE

CONCLUDING REMARKS

The present book offers the semi-diplomatic transcription of a late Middle English version of Constantinus Africanus' *Venerabilis Anatomia*, a manuscript housed in London, Wellcome Library, MS Wellcome 290 (ff. 1r-41v). Apart from the edition itself, the book contains other chapters concerned with different but related topics, which no doubt will provide the reader with a detailed outline of the production of manuscripts in mediaeval England together with the language of the period. In addition, information about the editorial principles and the compilation of the glossary is also included.

Chapter one has been concerned with the source and contents of the manuscript under scrutiny, providing authorship attribution and identifying an existing copy of the same manuscript, which is also housed in London, Wellcome Library (MS Wellcome 397, ff. 1r-14v). Furthermore, this chapter has made a codicological (material, dimension, decoration, ink, quiring, ruling and foliation) and palaeographic description (script, numerals, marginalia, catchwords, abbreviations, manuscript corrections and punctuation). The evidence obtained helps corroborate the middle of the 15th century as the likely date of composition for the manuscript.

Chapter two, in turn, has focused on the language in W290. This is a key section when it comes to Middle English texts insofar as it presents a high level of dialectal variation. Consequently, W290 allows for a reliable linguistic analysis as it contains more than 10,000 items, which constitutes an adequate sample of the language of the period. For the purpose, the main word classes (nouns, adjectives, personal pronouns, adverbs and verbs) have been described, offering frequencies when variation was found. Finally, the dialectal provenance of W290 has been provided using the methodology in *A Linguistic Atlas of Late Middle English* (McIntosh et al. 1986), which has demonstrated that the text in W290 was written somewhere to the north of river Thames in the county of Essex, nearby the city of Chelmsford, thus following the conventions of the East Midlands dialect.

Chapter three contains the editorial conventions together with the semi-diplomatic edition of the witness. First, the decisions that were made in terms of spelling, capitalization, word division, punctuation and abbreviations, among others, are explained in detail. Second, the semi-diplomatic edition is offered, in which line numbers, references to the folios and two apparatuses (for the marginalia and for manuscript corrections, respectively) are included.

Finally, chapter four offers a complete glossary of the words in W290, in which each entry contains the following information: headword, part of speech, meaning and the different allographs for each item with their corresponding number of occurrences. The glossary is of paramount importance in the edition of mediaeval *Fachprosa*, as it will help the reader understand the contents of the text, making the present work relevant not only for fields such as Historical Linguistics or Palaeography, but also to the History of Medicine or Surgery, among others.

References

Arakelian, Paul G. "Punctuation in a Late Middle English Manuscript." *Neuphilologische Mitteilungen* 76 (1975): 614-624.

Barber, Charles, Joan C. Beal and Philip A. Shaw. *The English Language: A Historical Introduction*. Cambridge: Cambridge University Press, 2009.

Baugh, Albert C. *A History of the English Language*. London: Routledge, 1959.

Benskin, Michael. "The letters <þ> and <y> in Later Middle English, and Some Related Matters." *Journal of the Society of Archivists* 7 (1982): 13-30.

Benskin, Michael. "The 'fit technique' explained." In *Regionalism in Late Medieval Manuscripts and Texts*, edited by Felicity Riddy, 9-26. Cambridge: D. S. Brewer, 1991.

Blake, Norman. Introduction to *The Cambridge History of the English Language* (Vol. 2, 1066-1476), edited by Norman Blake, 1-21. Cambridge: Cambridge University Press, 1992.

Brown, Michelle. *A Guide to Western Historical Scripts from Antiquity to 1600*. London: The British Library, 1990.

Brunner, Karl. *An Outline of Middle English Grammar*. Oxford: Basil Blackwell, 1970.

Burrow, John and Thorlac Turville-Petre. *A Book of Middle English*. Oxford: Blackwell Publishers, 1996.

Calle-Martín, Javier and Antonio Miranda-García. "Editing Middle English Punctuation. The Case of MS Egerton 2622 (ff. 136-152)." *International Journal of English Studies* 5.2 (2005): 27-44.

Calle-Martín, Javier and Antonio Miranda-García. "The Punctuation System Of Elizabethan Legal Documents: The Case of G.U.L. MS Hunter 3 (S.1.3)." *Review of English Studies* 59 (2008): 356-378.

Calle-Martín, Javier and Antonio Miranda-García. *The Middle English Version of De Viribus Herbarum (G. U. L. MS Hunter 497, ff. 1r-92r). Edition and Philological Study*. Bern: Peter Lang, 2012.

Calle-Martín, Javier and David Moreno-Olalla. "Body of evidence: Middle English annotated corpora and dialect atlases." In *Studies in English Medieval Language and Literature, Volume 38: The Use and Development of Middle English: Proceedings of the Sixth International*

Conference on Middle English, Cambridge 2008, edited by Richard Dance and Laura Wright, 17-33. Frankfurt and Main: Peter Lang, 2008.

Cappelli, Adriano. *Dizionario di Abbreviature Latine ed Italiane*. Milan: Ulrico Hoepli, [1912] 1990.

Crespo-García, Begoña and Emma Lezcano-González. "A preliminary Approach to dialectal localisation in Middle English." In *Interpretations of English: essays on language, linguistics and philology*, edited by Isabel Moskowich-Spiegel Fandiño, 41-60. La Coruña: Universidad de la Coruña, 1999.

De Hamel, Christopher. *Medieval Craftsmen: Scribes and Illuminators*. London: British Museum Press, 1992.

De Weever, Jacqueline. *Chaucer Name Dictionary: A Guide to Astrological, Biblical, Historical, Literary and Mythological Names in the Works of Geoffrey Chaucer*. New York: Routledge, 2013.

Denholm-Young, Noël. *Handwriting in England and Wales*. Cardiff: University of Wales Press, 1964.

Derolez, Albert. *The Palaeography of Gothic Manuscript Books. From the Twelfth to the Early Sixteenth Century*. Cambridge: Cambridge University Press, 2003.

Fischer, Olga. "Syntax." In *The Cambridge History of the English Language* (Vol. 2, 1066-1476), edited by Norman Blake, 207-398. Cambridge: Cambridge University Press, 1992.

Fischer, Olga and William van der Wurff. "Syntax." In *A History of the English Language*, edited by Richard Hogg and David Denison, 109-198. Cambridge: Cambridge University Press, 2006.

Fisiak, Jacek. "English Dialects in the Fifteenth Century: Some Observations Concerning the Shift of Isogloses." *Folia Linguistica Historica* 4.2 (1983): 195-217.

Hanna, Ralph. *London Literature, 1300-1380*. Cambridge: Cambridge University Press, 2005.

Hector, Leonard Charles. *The Handwriting of English Documents*. London: Edward Arnold, 1958.

Hogg, Richard M. and David Denison. *A History of the English Language*. Cambridge: Cambridge University Press, 2006.

Jekinson, Hilary. "Notes on the Study of English Punctuation of the Sixteenth Century." *Review of English Studies* 2 (1926): 152:158.

Keiser, George R. *A Manual of the Writings in Middle English 1050-1500. Volume X. Works of Science and Information.* New Haven, Connecticut: The Connecticut Academy of Arts and Sciences, 1998.

Kelly, Kate. *The History of Medicine: The Middle Ages*. New York: Infobase Publishing, 2009.

Laing, Margaret. "Multidimensionality: Time, Space and Stratigraphy in Historical Dialectology." In *Methods and Data in English Historical Dialectology*, edited by Marina Dossena and Roger Lass, 49-96. Bern: Peter Lang, 2004.

Lass, Roger. "Phonology and Morphology." In *The Cambridge History of the English Language* (vol. 2, 1066-1476), edited by Norman Blake, 23-154. Cambridge: Cambridge University Press, 1992.

Lass, Roger. "Phonology and Morphology." In *The Cambridge History of the English Language* (vol. 3, 1476-1776), edited by Roger Lass, 56-186. Cambridge: Cambridge University Press, 1999.

Lewis, Robert E. et al., eds. *Middle English Dictionary*. Ann Arbor: University of Michigan Press, 1952-2001. Online version in *Middle English Compendium*, edited by Frances McSparran. Ann Arbor: University of Michigan Press, 2000-. Available from http://quod.lib.umich.edu/m/med/

McIntosh, Angus, Michael Louis Samuels and Michael Benskin. *A Linguistic Atlas of Late Medieval English*, 4 vols. Aberdeen: Aberdeen University Press, 1986.

McIntosh, Angus. "The analysis of written Middle English." In *Middle English Dialectology: Essays on Some Principles and Problems*, edited by Margaret Laing, 1-21. Aberdeen: Aberdeen University Press, 1989a.

McIntosh, Angus. "A new Approach to Middle English Dialectology." In *Middle English Dialectology: Essays on Some Principles and Problems*, edited by Margaret Laing, 22-31. Aberdeen: Aberdeen University Press, 1989b.

Milroy, James. "Middle English Dialectology." In *The Cambridge History of the English Language* (vol. 2, 1066-1476), edited by Norman Blake, 156-204. Cambridge: Cambridge University Press, 1992.

Miranda-García, Antonio and Joaquín Garrido-Garrido. *Text Search Engine*. Málaga: University of Málaga, 2013.

Moorat, Samuel A. J. *Catalogue of Western Manuscripts on Medicine and Science in the Wellcome Historical Medical Library Vol. I. Manuscripts Written Before 1650 AD*. London: Publications of the Wellcome Historical Medical Library. Moore, Samuel, Sanford B. Meech and Harold Whitehall. 1935. "Middle English Dialect Characteristics and Dialect Boundaries." In *Essays and Studies in English and Comparative Literature*, 13: 1-60. Michigan: University of Michigan Publications in Language and Literature, 1935.

Mossé, Fernand. *A Handbook of Middle English*. Baltimore: The Johns Hopkins Press, 1952.

Newton, Francis. "Constantine the African and Monte Cassino: New Elements and the Text of the Isagoge." In *Constantine the African and 'Alī Ibn Al-'Abbās Al-Mağūsī: The Pantegni and Related Texts*, edited by Charles Burnett and Danielle Jacquart, 16-47. Leiden, New York and Koln: E.J. Brill, 1994.

Oakden, J. P. *Alliterative Poetry in Middle English: the Dialectical and Metrical Survey*. Manchester: Manchester University Press, 1930.

Oxford English Dictionary. CD-ROM Version 3.1. Oxford: Oxford University Press, 1962.

Pahta, Päivi and Irma Taavitsainen. "Vernacularisation of scientific and medical writing in its sociohistorical context." In *Medical and Scientific Writing in Late Medieval English*, edited by Irma Taavitsainen and Päivi Pahta, 1-22. Cambridge: Cambridge University Press, 2004.

Parkes, Malcolm Beckwith. *English Cursive Book Hands, 1250-1500*. London: Oxford University Press, 1969.

Parkes, Malcolm Beckwith. *Pause and Effect. An Introduction to the History of Punctuation in the West*. Hans: Scolar Press, 1992.

Petti, Anthony Gaetano. *English Literary Hands from Chaucer to Dryden*. Cambridge, Mass: Harvard University Press, 1977.

Prioreschi, Plinio. *A History of Medicine: Medieval Medicine*. Omaha: Horatius Press, 1996.

Roberts, Jane. *Guide to Scripts Used in English Writings up to 1500*. London: the British Library, 2005.

Romero-Barranco, Jesús. "A Semi-diplomatic Edition of MS Wellcome 290 (ff. 42r-48v)." *Analecta Malacitana* 35 (2012): 275-295.

Roura Javier, María Isabel. "The Present Participle in Middle English: the Establishment of the ending -ng." In *Interpretations of English. Essays on Language, Linguistics and Philology*, edited by Isabel Moskowich Spiegel-Fandiño, 207-214. La Coruña: University of La Coruña, 1999.

Samuels, Michael Louis. "Some Applications of Middle English Dialectology." In *Middle English Dialectology: Essays on Some Principles and Problems*, edited by Margaret Laing, 64-80. Aberdeen: Aberdeen University Press, 1989.

Smith, Jeremy J. *Essentials of Early English*. London: Routledge, 1999.

Tannenbaum, Samuel A. *The Handwriting of the Renaissance*. New York: Columbia University Press, 1930.

Talbot, Charles H. *Medicine in Medieval England*, Oldbourne: Londres, 1967.

Van Gelderen, Elly. *A History of the English Language*. Amsterdam: John Benjamins Publishing Company, 2006.

Voigts, Linda. "Medical prose." In *Middle English prose: A critical guide to major authors and genres*, edited by A.S.G. Edwards, 315-335. New Brunswick, NJ: Rutgers University Press, 1986.

Voigts, Linda Ehrsam and Patricia Deery Kurtz. *Scientific and Medical Writings in Old and Middle English. An Electronic Reference*. Ann Arbor: Michigan University Press, 2000.

Notes

[1] Mossé (1952: 2) terms them Kentish dialect (South-East), Southern dialect (South-West), East Midlands dialect, West Midlands dialect and Northern dialect.
[2] In addition to this treatise which opens the volume, there is also a short Middle English version of Galen´s Anatomy in ff. 42r-56v (Romero-Barranco 2012).
[3] By the year 1475 the first phase of the vernacularization of scientific and medical writing was largely complete, to such an extent that the use of the vernacular became even more common than Latin for the rendering of scientific material (Voigts 1986: 316; Pahta & Taavitsainen 2004: 12).
[4] According to Moorat (1962: 267), the treatise is written in vellum with a 19th-century calf binding, it was written in a neat semi-current charter hand, 28 lines to a page, where the margins have been cropped in re-binding.
[5] During the thirteenth century, the Collège (or Confrérie) de Saint Côme ("College - or Fraternity - of St. Cosmas "), in Paris, was the establishment of a guild of surgeons under the patronage of Saint Cosmas. It seems that it was established by Philippe the Fair (1285-1314) in 1311 by a special commission. As for the admission, the college required an examination by a panel of chosen surgeons. As a result, it tended to create a surgical aristocracy (*chirurgiens de la longue robe*, "Surgeons of the long robe") who held a position half-way between learned physicians on one side and barbers (who could not wear a long robe) on the other. The function of the guild was twofold: to obtain more respect and privileges for surgeons and to prevent encroachment into the surgical field by barbers (who were supposed to practice only minor surgery). In 1364, a royal statute banned illicit surgical practice in the city and proceeded to confirm the right of the College of St. Cosmas to examine and license surgeons together with the royal surgeons and the provost of Paris. The College of St. Cosmas continued to exist until the French revolution (Prioreschi 1996:430).
[6] All treatises except the second and the fifth are divided into *doctrinae*, and each *doctrina* into chapters (Prioreschi 1996: 458).
[7] From head to foot.
[8] Calfskin (*carta vitulina*, vellum) was the largest [kind of parchment] in size and had an even sturdy surface, its hair side and flesh side having approximately the same white colour. Apart from calfskin or vellum, there were other writing surfaces such as sheepskin or goatskin, which differed in colour (Derolez 2003: 31). In addition, there was an especially fine variety, known as 'abortive' or 'uterine' vellum and normally used for the more sumptuous literary and liturgical manuscripts. It was made (or was said to be) made from the skins of stillborn and unborn calves, kids and lambs (Hector 1958: 15).
[9] According to De Hamel (1992: 32), there were "a fair number of medieval recipes for making ink", from which two different kinds could be highlighted: carbon ink and metal-gall ink. The former is made of charcoal or lamp-black mixed with a gum, while the latter mainly composed of iron gall, made by mixing a solution of tannic acids with ferrous sulphate (copperas).

[10] In the date of composition of the present manuscript, ruling was invariably traced on the hair side of the parchment, so that the ridges appear on the flesh side (Derolez 2003: 35).
[11] This mark did not always stand for *us*, a *q* having this graph after it stood for the enclitic *que* as well as for the words *quod*, *quia*, and even *quam* (Tannenbaum 1930: 127).
[12] According to van Gelderen (2006: 111), "this change occurs at different times in different parts of Britain: in the North and East, it proceeds faster than in the South and West, probably due to Scandinavian influence".
[13] There are now [in Middle English] two noun cases, genitive and 'common case'. Virtually all nouns have gone to the *s*-genitive and plurals, except for weak nouns that retain *-en* (Hogg and Denison 2006: 71).
[14] Comparatives and superlatives have been disregarded in this table.
[15] The adjective *hard* in the comparative can be either *harder* or *herder*. The other two forms, however, always change the stem vowel.
[16] No possessive pronouns seem to be found in W290.
[17] Irregular weak verbs like bringen > broughte have been disregarded in this table.
[18] The aim of dialectology is to take into account three analytical planes of language: a) how linguistic forms change through time (diachronic); b) how they vary accross space (diatopic); and c) how the interactions of the speakers and writers of the language produce and define this variation (diastratic) (Laing 2004: 49).
[19] 22 literary texts out of 44, and 240 non-literary documents out of 266 are from the time after 1400 (Fisiak 1983: 200).
[20] It needs to be noted that, for many academics, *LALME* is not elsewhere taken as a reliable source for Middle English dialectology, and that is the reason why linguistic analyses become of paramount importance in studies of this nature. The arguments offered by scholars range from the unclear dialectal boundaries in the *Atlas* (Calle-Martín and Moreno-Olalla 2008: 24) to the systematic ascription of texts to the county of Essex, as its language constituted "a perfectly acceptable mode for professional London writing in the fifteenth century" (Hanna 2005: 28).
[21] Even though the analysis that has been carried out leads us to state that W290 seems to have been written in the county of Essex, one can never be completely sure whether the scribe was actually a native speaker of that zone or, on the contrary, whether he acquired his habits of written language therein (Benskin 1982: 14; Crespo-García and Lezcano-González 1999: 48).